(Un)explainable Technology

Hendrik Kempt

(Un)explainable Technology

palgrave
macmillan

Hendrik Kempt
Applied Ethics Group
RWTH Aachen University
Aachen, Germany

ISBN 978-3-031-68097-7   ISBN 978-3-031-68098-4   (eBook)
https://doi.org/10.1007/978-3-031-68098-4

Cover illustration: © Melisa Hasan

This Palgrave Macmillan imprint is published by the registered company Springer Nature Switzerland AG
The registered company address is: Gewerbestrasse 11, 6330 Cham, Switzerland

If disposing of this product, please recycle the paper.

For Telmen

ACKNOWLEDGMENTS

This book project started as a continuation and summarization of my PhD thesis in which I developed a pragmatic perspective on the role of explainability in medical decision-making. However, as I noticed how this topic cannot be fully appreciated without addressing the role of explainability in our interactions with technology at large, the project grew to a more general account of explainability.

In developing the core ideas of my work, I have had the chance to talk to many different people about my thoughts and was encouraged and inspired by their time and patience, their useful suggestions, their cooperation. For this, I am indebted to many people and extend my gratitude to all of them. First and foremost, I would like to thank Saskia K. Nagel for providing me with the opportunity, freedom, and advice to pursue my PhD thesis under special circumstances. Without her immense support, I would not be working in philosophy anymore. Additionally, I would like to thank her and Sven Nyholm, who co-supervised my PhD thesis, for the feedback and valuable insight, which helped me develop some of the thoughts I wrote down in this book.

My colleagues, former colleagues, and co-authors Jan-Christoph Heilinger, Nils Freyer, W. Jared Parmer, Niel Conradie, Camilla Colombo, Giacomo Figá-Talamanca, Frieder Bögner, Peter Königs, Chae-Won Yun, Niklas Michels, and Michelle Görlitz. They all have helped in their particular ways to keep this project afloat: by providing encouragement or feedback, co-authoring some of the papers, challenging my arguments, or simply by being good company. I owe all of them my sincere thanks, and hope to be able to provide the same to them in the future. Of course, without the necessary but often invisible and unpaid labor of anonymous reviewers in peer-reviewed journals, this book would have not been possible as well. Their feedback has improved the basis of my argument in many instances, and I owe them thanks.

I also extend my gratitude to Rachael Ballard and Naveen Dass from Palgrave MacMillan, who guided the process with a level of professionalism, support, and understanding that I have come to value highly.

Lastly, I want to thank my partner Telmen for teaching me the joy of the unexplainable.

CONTENTS

About the Author

Hendrik Kempt is a Postdoctoral Researcher at the Applied Ethics Group, RWTH Aachen University, where he received his Ph.D. in 2023. His work focuses mainly on questions of the ethics of human-machine interaction, of artificial intelligence and other emerging technologies, and of technology in general.

He has previously published the monographs "Chatbots and the Domestication of AI" (2020) as well as "Synthetic Friends—A Philosophy of Human-Machine Friendship" (2022), edited a book with Megan Volpert on "RuPaul's Drag Race and Philosophy", and has written a number of journal articles and book chapters on issues regarding the ethics of technology.

CHAPTER 1

Introduction

Abstract This chapter introduces the core philosophical questions of explainability and the innate human quest for explanations. We first critically assess the metaphor of AI as a black box and discuss how much we actually know about the functioning of AI. In entertaining a thought experiment about an unfailing oracle and the scrutiny it would receive today, we uncover practical and epistemic purposes for explanations, and transfer those to explanations of technological behavior. This way, we gather elements of what would make a good explanation for technology, and we set out the task for their epistemological and ethical questions.

Keywords Explainability · Black Box AI · Granularity · Parsimony

Explanations are practically useful for a myriad of reasons. Good explanations make things more interactable, controllable, measurable, foreseeable, predictable, and even comfortable. All these features give humanity an increased range of agency and an increased sense of security. Explanations were used to enchant and disenchant the world by positing either the presence of supernatural entities or the absence of anything supernatural. Both strategies helped to gain a however subjective control over the world by learning what to expect from it.

Explanations for events in or facts of the world usually require a causal chain of events. An event happens for the causal reasons x, y, z and would have been counterfactually not happening if causal reasons were different or absent. Different theories of causality help to provide explanations for these kinds of events and facts. Some events in the world are actions, which are based on decisions. Causal explanations hardly ever provide insight here: describing human actions and decisions *causally* may be useful in some minor, specified contexts, but usually, we require *normative* explanations that elaborate on the reasons why someone is doing or did some action.

At this stage, it is well understood that we have reached the limits of causal explanations for contemporary AI, which appears to some a genuinely new problem for the ethics of technology. We can seemingly ask some AI, in the same way we do for human agents, what its reasons are for providing some output, but we still seek causal explanations for its behavior to be considered acceptable in a normative sense. This is because barely any of our blame-assigning practices, our responsibility practices, or even our intuitions about what these practices should be based on (i.e., a certain theory of mind of machines that currently most people would not assign to any available AI) fit the phenomenology of dealing with these machines. Other AI-driven machines do not even grant us that opportunity but also have become virtually ubiquitous but invisible parts of our daily lives.

Most machines we have managed to design and build in the recent past were largely based on well-understood principles of mechanics, at least to the degree that we could predict how those machines would behave. It was, actually, necessary to understand these principles to create machines that would adhere to these principles reliably, and not break down or go "off script". If we did not know these principles to a level of minimal tolerance, most machines would not have seen the light of day.

With the advent of principally unexplainable artificial intelligence, it appears that we have returned to some stage of unknowing or self-inflicted lack of explanations for the specific behaviors of some events in the world. The worries reemerge that we are losing control, or at least becoming more exposed to the unpredictability of algorithms, their well-known but poorly controlled biases, and the uncomfortable contingency of failure of artificially driven decisions.

These issues do transfer to institutional decision-making such as bureaucracy, legislation (think of bureaucratized decision-making procedures in hiring, judging, apartment renting, loans and credit scores, etc.): we want these decisions not only to be "correct" (in the sense that they are equitable) but also explainable: we want to their explanations to be based on known parameters, so they are somewhat predictable (for those with knowledge of the relevant parameters). Philosophers and engineers alike have taken up the task of engaging this fact, to find strategies to mitigate the consequences of unexplainable technology and minimize the lack of knowledge we currently have. Such a research project appears clearly motivated and normatively justified.

Being subjected to decisions that appear unexplainable to us, is something that happens to us—upon closer inspection—all the time. Be it because the decision-making process is opaque or even secret, or because other people deny us to elaborate on their reasons to act in certain ways, or because we lack the capacity or capability to understand the reasons why something is happening the way it is happening. We rely on experts to explain the world to us in terminology that those very experts will admit is highly reductive and under-complex, and yet, all these things are fine. The very fact that we are already subject to some algorithmic decisions without being aware or especially bothered by it is another hint that things might not be as dire as I just portrayed them.

In this book, I am aiming to resolve the tension between the issues of unexplainable decisions (especially those made by autonomous technologies), and the fact that we in other, even non-tech contexts, do not seem to mind being made subjects to decisions for which explanations are not readily available.

1.1 STRUCTURE OF THE BOOK

This book is structured into several chapters exploring the different dimensions and practical aspects of the question of unexplainable technology. The first chapter concerns the struggle for an explanation in technology in the first place. As an interdisciplinary project, the search for shared epistemic and practical purposes that dictate the depth and granularity of explanations is a particularly hard task. This search is often complicated by conceptual and terminological differences in disciplines that often work independently from each other. The translational work needed here is thus substantial. In Chapter 2, such a translational attempt is undertaken by proposing a conceptual and terminological setup that aims at catching many different traditions of structuring these debates. From larger distinctions of explainability from interpretability, for example, to sub-distinction of these concepts, and practically oriented demands of contestability and auditability, we aim to clarify the issue emerging in an interdisciplinary field that is using an array of different terms. Turning to Chapter 3, the first "harvest" of Chapter 2 can be found: having clarified some of the terminology, the epistemological conditions for when a technology can count as "explained", can be formulated. This also demonstrates how the epistemology of AI and the ethics of AI treat explainability rather differently, as the epistemic purposes of explainability do not necessarily bear on whether AI *should* be explainable. Chapter 4, consequently, approaches the normative issue of explainability of technology. After discussing the many different instrumental reasons for pursuing explainability (all of which, however, can be weakened/avoided), we are facing one of the strongest arguments in favor: the right to explanation. In analyzing the implications of such a right, and rejecting it based on the function of expertise in society, we arrive at a *claim to expertise* we have in dealing with unexplainable decision pathways, rather than a right to explanation. We then turn our attention to the political ramifications of regulating unexplainable technology. In Chapter 5, we approach four distinct areas in which unexplainable technology is present and can be discussed addressing the distinctions and arguments brought forth so far: medical procedures, self-driving cars and autonomous weapons systems, generative AI and large language models, and recommender algorithms. These represent a diverse chunk of AI-applications with immediate impact on our lives. In Chapter 6, we summarize the arguments from previous chapters and attempt to sharpen the thesis towards the question of whether explainability, besides its initial appeal and its foundation in the human condition, amounts to an ethically relevant value if all other requirements are fulfilled. We will see that explainability is, in fact, not worth pursuing in and of itself. Lastly, in Chapter 7, we will close out and remind ourselves that the demand for strong explainability requirements is debunked, but the pursuit still is useful in some areas.

1.2   On the Metaphor of AI as a Black Box

The metaphor of AI as "black boxes" is popular among scientists, philosophers, and the wider audience alike (see, e.g., Castelvecchi, 2016; Pasquale, 2015), even with those who usually prefer precise language (e.g., Wachter et al., 2018, or Zednik, 2019). It is a popular one due to the forcefulness of its imagery and its supposed solution (making said box transparent to become a "glass box"). The depiction of AI as a black box, however, also creates an image of AI as being fully unknown, as do some other comparisons (like the one we will discuss in this chapter). The idea is that the decisional pathways of an AI are as unknown as can be, and thus we can only observe the input and the output, but not how these compare to each other, or how the input bears on the output.

We ought to take a more realistic perspective here, to also get a better grip on what the issue with unexplainable technology actually is. The realistic perspective requires us to acknowledge that we do possess quite a bit of knowledge about the construction, and even inner workings of AI. The following only illustrates some areas of constructing and training AI (rather superficially, admittedly—we do know much more) in which there is secured knowledge.

1.2.1   Hardware and Software Principles

Some of the processes with which we design artificial intelligence algorithms are well understood. From creating convolutional neural networks and transformer models that manage to find and reproduce patterns, to the specific architectures of energy-efficient and self-optimizing processors, the ability to build the software and hardware of these machines lies well within the know-how domain of humanity. Even while some of the research and development of the latest processors are done by AI, and especially the robotic creation of the hardware (which requires an otherwise impossible amount of precision), the question of explainability never emerges on that level. The material and developmental processes and requirements are gaplessly understood. Simply put: the fact that we can design something like a "black box" already means that it cannot be a fully black box.

1.2.2   Data

One of the key insights that changed artificial intelligence from "good old fashioned AI" (GOFAI, Haugeland, 1985) to machine learning was that the progress in performance achieved by training algorithms on data scales reliably with the amount and quality of data used. This means that the assembly and curation of data sets have become a key "ingredient" in creating AI of a certain sophistication. The better the training data, the better the results in the final model. We can surmise from the output of some models whether the underlying training data was corrupted, faulty, or otherwise problematic by the quality of said output. This does not determine any specific output, but it suggests that the selection of the kinds of data going into the training of an AI will influence the range and quality of the output it will produce. We know how to improve data sets, and what kind of data works best, to train algorithms for improved performance.

1.2.3   Nodes and Weights

At this stage, even with incredibly large models, we understand how their functions process the input data, even if this does not allow for determinant conclusions about the output. We also understand rather well how we can improve the output of a machine through fine-tuning, which affects the weights within a given model. RLHF, a state-of-the-art method of tuning the models for a certain kind of desired output (Sun et al., 2023), is being successfully used to train chatbots on previously impossible levels of sophistication in the groundedness, creativity, context-appropriateness, and reasoning. While the causes for specific decision pathways between input and output remain unclear, the fact that there are decision pathways of a certain length and depth (i.e., weightedness) that we can influence in a rule-governed way proves that the processes are not fully unknown, either.

That being said, however, it is still true that many of the failings and mistakes, as well as their corrections, can only be identified post-hoc, fixed post-hoc, and explained post-hoc. These kinds of "post-hoc rationalization"—explanations count as a subset of interpretability explanations of current AI technology because they do not provide explanations about the processes themselves, but retro-fit the outputs to the inputs based on the knowledge available about the statics of the neural network in question.

This is because the level of explanations needed is usually for single instances of behaviors of an AI model. The required explanation, thus, is not achieved by reference to the nodes and weight and training data used, but by the specific run of the decision pathway through the nodes in that very instance. However, since the training makes the weights virtually impossible to reconstruct, and the amount of nodes and connections virtually impossible to count, explaining why in that one instance the AI did that one thing is, virtually, impossible.

As stated before, higher-level explanations of the same instance are available, however. This creates the question of how we should characterize AI at this stage, viewed from a medium distance with practical demands in public discourse. Ultimately, calling it a "black box" is suggestive of less knowledge than we actually possess and might influence the perception of the lay audience beyond its due. On the other hand, it is not a glass box in the sense that we can observe and meaningfully control the inner workings of a machine. It would be, thus, more apt to speak of a gray box: it is not a black box in the sense that there is no insight, and it is not a glassy one in which there is full clarity of the causes of the inner workings. Whether or not the gray box is sufficiently transparent remains to be shown in the following chapters, in which we explore the epistemology of explainability, the ethical requirements, and some applied cases. Especially going through some of the applied cases demonstrates the differing relevance of explainability altogether.

Since the metaphors are deployed to help understand an otherwise complex issue with more clarity, and none of these appear to do so, we should avoid talking about AI and other technologies as any kind of box (unless, of course, the technology in question is a literal box).

1.3 THE UNFAILING ORACLE AND OUR RELATIONSHIP WITH UNEXPLAINABLE DECISIONS

Suppose, just for a moment, that the Oracle of Delphi would still exist today, but in even better shape than before—i.e., not only with less semantic ambiguity in its predictions but also with a lower fail rate (a similar thought experiment is entertained in Armstrong et al. [2012], but with different intuitions). The process, however, is still the same: shrouded in mystery and hermetic practices, the oracle receives answers to hopeful visitors' problems through not further explained methods. It is unlikely that this oracle would last long before it would be questioned, put to the test, and demanded to make predictions in controlled circumstances. I do not doubt that even if the oracle held up such scrutiny, testing, and explanation-seeking would never cease. This, I suspect, is for two reasons: on the one side, we would like to know how we can replicate the oracle's success to use it elsewhere. Oracles with high predictive skills are a valuable asset. However, on the other side, we also simply want to understand in a way that mirrors or integrates into our scientific worldview. There are gaps in understanding between asking the oracle for advice and receiving useful, precise answers that we could not have gotten elsewhere.

Maybe this also explains how ways of interpreting and predicting the world that do not mesh with the scientific method as a baseline for acceptable explanations, still require explanations that are "causal": astrology, if done "correctly", has meticulous tables that relate one's birth date and time to all sorts of influences of astronomical nature; homeopathy provides a story about the workings of their products; even alchemy had "causal" explanations which, ultimately, just turned out to be false. In short, even pseudo-sciences require a certain kind of gaplessness in their explanations, as they would otherwise fall victim to the human demand for understanding.

Most activities, even the divinely inspired or guaranteed ones, require an explanation for humans to accept them in the long run. The growing, if not overwhelming, success of the scientific method as a premise for any acceptable causal explanation in most contexts of inquiry, however, shows that these explanations are not merely a cognitive requirement: they also have to have roots in utility, and the scientific method is the most successful in this regard: causal explanations that are based on testable, empirically supported, reproducible experiments have led to the most stable predictions available.

1.3.1   The Phenomenology of Seeking Explanations

We are constantly confronted with other people's unexplained decisions occurring in our lives, and often enough we do not care to ask. This is for a myriad of reasons, and yet most of them are practical: as many of these explanations would not help us in our daily lives, even if we are partially affected by them.

However, there are also strong cognitive reasons to merely assume other people's reasons for their decisions. Often enough, they are uninformative, as their reasoning is banal, highly subjective, or even erratic without external insight. However, it is also clearly overwhelming to try to inquire with every person who is influencing our lives what explanation they would give for their decisions. Simply put, in many instances, we do not ask for explanations.

However, as we have seen with the example of the oracle, it appears that the fact that some decisions could not be explained even if we asked for an explanation seems to change our attitude towards those decisions. Once the potential for an (easy) explanation is gone, we are incensed to inquire even more. There are certain dimensions of the eeriness of unexplained and unexplainable phenomena, a space in our mental cartography of the world, the need for completeness of the scientific worldview, and other higher-level reasons why unexplained and unexplainable decisions are not the same.

Seeking explanations might simply be best understood as part of the human condition: something humanity will always strive for, independent of externally motivating reasons like practical purposes, but for the mere sake of understanding. This does not guarantee that we always seek true explanations, as we see with pseudo-science, fake news, confirmation biases, and other elements of cognitive maladaptation or ignorance. This also does not guarantee that we seek even complete explanations: often enough, practical considerations override our innate quest for understanding and we are satisfied with explanations that suffice for our practical purposes, whether they are true, complete, or even plausible. Sometimes, an explanation is enough.

In this vein, we should understand the quest for technological explanations as motivated by innate human curiosity, on the one side, but also by practical and cognitive purposes, on the other side. However, we should seek good explanations and must clarify what a good explanation for a given technology means. In the next chapter, we will elaborate on some of the requirements for such a good explanation.

1.4   WHAT MAKES A GOOD
EXPLANATION IN TECHNOLOGY

In this section, we will elaborate on some elements required for a good explanation of the behavior of autonomous technologies. As we have pointed out before, the human condition to seek explanations has a positive side, which makes us intellectually curious and persistent in our quest to understand the world. However, this very condition to demand an explanation can on occasion make us vulnerable to seek *any* explanation, or remain committed to bad ones: objectively wrong ones; ones biased by our own prejudices and merely confirming what we already believe to "know"; pseudo-scientific ones which make claims about the world that cannot be proven; ones that have been demonstrated to be false; ones we make up for our own peace of mind without trying to reflect the evidence; and so on.

As we seek explanations that explain a shared phenomenon—the functioning and behavior of autonomous machines—our explanations should also be in such form so that we can share them as we want to gain a shared understanding. This means that the reasons we provide for their behavior should be based on intersubjective or objective reasons. To achieve this, we should ensure that the theories that provide us these reasons, and the evidence we produce for certain explanations, are of a certain quality.

1.4.1   *Explanatory, Descriptive, and Predictive Power of Theories and the Intentional Stance*

With the "explanatory power of a theory", philosophers of science usually mean that a theory can provide a cohesive account of why something happened, i.e., it can explain past events. The more events can be explained with one theory only, the more explanatory power it has. The predictive power of a theory means that we can use this theory to make predictions about future events. Both powers are important for picking a frame to explain certain events.

Take, for example, Daniel Dennett's conceptualization of the "intentional stance" (1981). This approach suggests that we decide to take a stance towards something *as if it had intentions,* even if we cannot be sure that something does have intentions in this sense (animals, AI, etc.). We do this because the explanations we can make in reference to intentions can increase the explanatory and predictive power of behavior by assuming it behaves as if it were a rational agent with certain mental states (i.e., intentions). We can observe this with dogs: if we characterize their behavior as guided by intentions, we might be able to explain some past and predict some future behaviors, which otherwise would not be explainable based on conditioning and other behavioral elements alone (see Byrne's analysis of the relevance of explanatory stances in Byrne [2023]).

The intentional stance is not necessarily akin to anthropomorphization, i.e., the projection of human qualities on non-human entities. While anthropomorphization can be an explanatory stance (especially noticeable in myths, but also in human interactions with animals and AI), it often is largely a psychological one, i.e., a stance taken for psychological, rather than actual explanatory reasons. These psychological reasons could include a certain de-mystification, complexity reduction, loneliness, etc. They are not even necessarily predicated on the unexplainedness (or unexplainability) of the technology in question: even well-explained, non-autonomously behaving entities are often being anthropomorphized. But without any way of understanding how the technology operates, the easiest way to find an explanation is to draw parallels to human (or animal) behavior. Anthropomorphizing, thus, is mostly not about the quality of an explanation but about the necessity of having an explanation. The intentional stance, however, describes how taking such a stance actually improves our explanatory and predicting capacities about a behavior.

Currently, the intentional stance towards autonomous technology is often rejected on methodological and practical grounds (although some argue for it, see Zednik [2019], and Zerilli [2022]). While for some purposes, the intentional stance is a sensible perspective to take to explain and predict machine behavior (Papagni & Koeszegi, 2021), it would change the current goal of explaining AI from a causal to an intentional framework. Even if the intentional stance does not amount to anthropomorphization, it still requires the (at least methodological) ascription of mental states. As it stands, the explanatory advantage of such stands does not outweigh the costs of pretending that an AI has mental states.

1.4.2 Non-causal Explanations

Unsurprisingly, many disciplines of scientific inquiry and a lot of everyday life explanations do not rely on causal explanations of natural events but have other frames of reference. Especially normative frames of explaining behavior and decisions do not refer to causal reasons. Take, for example, the explanation of a standing, elderly woman clearing her throat next to a younger, able-bodied man sitting on a bus. One could explain this in causal terms of her throat itching, but also in a highly socially and morally embedded context in which her intentionally clearing her throat and thus catching his attention would compel him to offer her his seat, as she would like to sit and thinks that the younger man should vacate his seat for her.

The reference to intentions, wishes and desires, purposes, goals, aims, and other motivating emotions, but also social and moral norms provides a wide range of explanations that have no causal ground in certain states or events. However, as we have seen with the intentional stance, non-causal explanations for machine behavior have, at this stage, merely a metaphorical or psychological ground (i.e., an explanation for the sake of the explaining person, not the explained subject). While this might change, especially if we should encounter emerging capabilities that could suggest intentional behavior, explaining behavior non-causally (or non-statistically, at least) might become a viable way. Since an "agential stance" comes with normative ramifications, however, the bar for such a shift towards normative or prudential explanations is set even higher than merely taking an intentional stance.

1.4.3   Explanatory Gaps and Granularity

Next to the question of what is even the right explanatory frame for explaining decisions made by technology, we should introduce the concept of the *granularity* of explanations (which is independent of the question of whether causal or non-causal explanations are the better approach). The answers will rely on the cognitive and practical purposes we pursue with the explanations given. However, many of these explanations, if not all, appear to adhere to one shared element that makes them comparable to each other even if the levels of explanatory depth are vastly different: they are gapless. Gaplessness here means that the explanations given are all within the same presupposed framework in which this explanation takes place (i.e., no "deus ex machina"—moment is allowed to skip some parts of an otherwise naturalistic explanation, while vice versa it would be noticeable if an explanation about the workings of God eventually referred to a natural law as one God has to obey, while elsewhere God's will is being put down as the main explanatory datum).

This may also mark the utility of explanations as outlined above: if they are gapless, they constitute a story (in the wider sense of the word) about an otherwise unintelligible event, state, or feature. Explanations at large reduce the complexity of the world by putting them into relation to each other, and providing human beings a sense of conquerability or at least the ability to assert oneself about an otherwise hostile world.

However, the paradigmatic case of a gapless explanation for most (natural) events in the world appears to be the causal explanation based on evidence. For normative explanations, e.g., of an action, we usually demand a clear if not deductive demonstration of normative reasons and options of action that explain the action of an agent that reduces, if not eliminates, questions we may have against the agent. Approximate, statistically likely, or otherwise non-determinant explanations are often considered helpful for a specific cognitive or practical purpose but appear to be insufficient or inappropriate for many explainability contexts.

1.4.4   Explanatory Coherence and Parsimony

Next to gaplessness and the need to determine a frame of reference for explanations, we also seek consistency and coherence in explanations so that they are in harmony with our other beliefs. This kind of cognitive harmony, in which explanations brought about are generally not implicitly or explicitly contradicting and undermining other explanations or beliefs elsewhere, is key for avoiding explanations that merely fit the explanatory demands. A good explanation is not only delivering any explanation but one that is embeddable into the larger picture of beliefs present.

Beliefs expressed in explanations ought to be in relative coherence with our other beliefs. The "deus ex machina"—reference from earlier demonstrates this clearly, as we ought to keep the presuppositions for explanations internally and externally free of contradictions. The idea of epistemic coherence, thus, refers to the relative absence of contradictions between existing beliefs and those beliefs produced by an explanation.

The principle of epistemic parsimony works in a similar fashion, usually known in the specific formulation of "Occam's razor": the fewer ontological or other assumptions have to be made to explain something, the closer to truth is that explanation. We ought only to assume what is necessary to explain a given explanandum, and explanations that have to assume less, are better. This is motivated by the possibility of constantly enriching (possibly at hoc) one's ontology to explain phenomena that are in coherence with previous beliefs (and thus are coherent), but merely because another layer is added to the explanation.

In the philosophy of technology, we are dealing with artifacts, and thus resort to naturalistic, causal explanations as the ontologically most parsimonious way of arguing. Introducing non-causal or even metaphysical explanations might thus unduly crowd our metaphysics of technological artifacts by adding metaphysical statuses to them to explain behaviors that we otherwise do not need to assume. We might be able to see (and explain) a deviation from this implicit explanatory frame with those computer scientists and researchers who argue some generative AI, most often highly complex chatbots like chatGPT (in the latest version), to have some consciousness. Introducing consciousness-claims (or other emerging capabilities) to these neural network-style artificial intelligences carries a lot of ontological claims that those who explain AI-behavior differently do not have to do.

Overstating the importance of epistemic harmony and parsimony could result in a confirmation bias, in which evidence is cherry-picked to fit a pre-established explanans, i.e., the explanandum is described in a way that the explanans gains the strongest explanatory power. This means that for "good explanations" to emerge, we also need an open-minded, precise, evidence-sensitive but not excluding account of describing the explanandum, the current state of secured knowledge surrounding the phenomenon to be explained, and the available explanations that fit these requirements and still provide some explanatory power.

1.4.5   Determining the Explanandum and the Explanans

This leads to the last, important, and yet almost trivial, element in good explanations: the semantic precision used both in determining the explanandum (what is to be explained) and the explanans (what is explaining). An ambiguous answer to the question of what is to be explained can make any competing explanations incommensurable, while ambiguity or lack of precision in the explanans can lead to an obscurum per obscurius or other unhelpful, inadequate explanations.

As we have seen in the Oracle case, semantic ambiguity was used to make predictions that could be interpreted both ways, generating no knowledge until after the predicted event clarified the ambiguity. An ambiguous explanation, thus, is virtually useless, as its descriptive and predictive powers are diminished. This shows that semantic precision and unambiguous terminology are preconditions for a good explanation.

In the question of unexplainable technology, we thus have to define what it is that is to be explained. First, we might point out that it is in the name: the explanandum is *the technology*. And once we specify what technology we have in mind, the explanations for it will be adequate. This, however, is of problematic semantic imprecision that different explanans might be in conflict about what the proper target of their explanation is: if "the AI" or some elements of it are the explanandum, what can be a sufficiently precise explanation in the first place?

We then might specify that, since we are dealing with autonomous technologies, their *behavior* is what is to be explained. However, while this offers a clearer picture of what may be the subject of an explanation, it still is unclear what behavior is meant here.

Lastly, thus, it is the *decision and the pathway* that lead to the decision of an autonomously behaving machine, such as an AI or AI-based robot. This is because the decision and the causal chain leading up to that decision, i.e., the reasons why a machine decided a certain way, are of the practical relevance that motivates an inquiry in the first place. It also matches best with the statistical approximation of an AI's decision-making process: as AI is classifying and reacting with specific "confidence intervals", i.e., levels of confidence of being correct about a classification or the like, the final decision is usually an aggregate of different confidence intervals and assessments. Explaining gaplessly how these form, however, is usually considered the issue at play.

REFERENCES

Armstrong, S., Sandberg, A., & Bostrom, N. (2012). Thinking inside the box: Controlling and using an oracle AI. *Minds & Machines, 22*(4), 299–324. https://doi.org/10.1007/s11023-012-9282-2

Byrne, R. M. (2023). Good explanations in Explainable Artificial Intelligence (XAI): Evidence from human explanatory reasoning. In *Proceedings of the thirty-second international joint conference on artificial intelligence*. International Joint Conferences on Artificial Intelligence Organization, Macau, SAR China (pp. 6536–6544).

Castelvecchi, B. (2016). *Can we open the black-box of AI?* https://www.nature.com/news/can-we-open-the-black-box-of-ai-1.20731 (last accessed May 31st 2024).

Dennett, D. C. (1981). True believers: The intentional strategy and why it works. In A. F. Heath (ed.), *Scientific Explanation: Papers Based on Herbert Spencer Lectures Given in the University of Oxford*. Clarendon Press. pp. 150–167.

Haugeland, J. (1985). *Artificial Intelligence: The Very Idea*. MIT Press.

Papagni, G., & Koeszegi, S. A. (2021). Pragmatic approach to the intentional stance semantic, empirical and ethical considerations for the design of artificial agents. *Minds & Machines, 31*, 505–534. https://doi.org/10.1007/s11023-021-09567-6

Pasquale, F. (2015). *The black box society: The secret algorithms that control money and information*. Harvard University Press

Sun, Z., Shen, S., Cao, S., Liu, H., Li, C., Shen, Y., Gan, C., Gui, L. Y., Wang, Y. X., Keutzer, K., & Darrell, T. (2023). *Aligning large multimodal models with factually augmented rlhf*. arXiv preprint arXiv:2309.14525

Wachter, S., Mittelstadt, B., & Russel, C. (2018). Counterfactual explanations without opening the black box: Automated decisions and the GDPR. *Harvard Journal of Law & Technology, 31*(2), 842–861. https://doi.org/10.2139/ssrn.3063289

Zednik, C. (2019). Solving the black box problem: A normative framework for explainable artificial intelligence. *arXiv*. https://arxiv.org/abs/1903.04361

Zerilli, J. (2022). Explaining machine learning decisions. *Philosophy of Science, 89*(1), 1–19. https://doi.org/10.1017/psa.2021.13

CHAPTER 2

Conceptual Clarifications

Abstract To tackle the issues present in the international and inter-disciplinary debate on explainability, we introduce and characterize the core concepts of explainability, interpretability, contestability, certifiability, auditability, post-hoc rationalizability, and transparency. In outlining the different epistemic and practical purposes pursued with these techniques, we can determine and disambiguate some of the confusion present in the debate, and clarify what it actually is we are interested in.

Keywords Explainability · Interpretability · Auditability · Transparency · Certifiability

Philosophers are notorious for making distinctions—which are often useful in clarifying a topic of concern. They are especially helpful when there are different purposes at play for the investigation of a topic.

For theoretical purposes, we might want to clarify concepts and their dependencies on each other (this is, generally speaking, the overarching goal of analytic philosophy). For practical reasons, we want conceptual reliability (which comes with conceptual clarity) to base normative arguments on them. That means that if we want to be able to debate about normative contents of applying a certain theory containing a specific concept, we want to do this error-free. An error, however, would emerge

© The Author(s), under exclusive license to Springer Nature Switzerland AG 2024
H. Kempt, *(Un)explainable Technology*,
https://doi.org/10.1007/978-3-031-68098-4_2

if our disagreement was hinging on different presuppositions about the concepts at play and their content. Conceptual disagreements can lead to defective discourses, and it is of practical relevance to share a common conceptual ground. Legal theory, e.g., relies heavily on defining concepts before making any kind of legal claim that contains these concepts. Next to conceptual errors, we may also encounter terminological errors, which can follow from ill-defined concepts, changing semantic traditions, a missing but relevant distinction, or a superficial characterization of the concepts at play: especially in everyday conversation terminological disagreements lead to fruitless conversation in which two people argue past each other due to homonymic uses of different concepts.

That means that a conceptual family may be described in different forms depending on the convincingness of the theoretical proposal or its practical utility. It is an open question in the methods of philosophy if we should prioritize their theoretical or practical utility. One can attempt to first find a convincing base definition of a concept and work our way down towards practical issues and use the defined concepts, or we can attempt to start with currently used meanings of a certain concept and reconstruct it by generalizing its contents to the degree that still retains the meaning as used in practice.

The question of what an explanation is has been keeping philosophers engaged for the past few hundred years as there are several approaches to that question even how an explanation relates to understanding and how interpretations feature in the deconstruction of epistemically or conceptually inaccessible explanations. In a rather obvious sense, an explanation leads to an understanding, and a good explanation leads to correct understanding, etc. However, the more closely we examine available definitions, whether they are explicit or implicit, we encounter a surprising lack of clarity.

That an interdisciplinary field such as explainable artificial intelligence (XAI) retains a certain incoherence in the definitions used is not surprising, and occasionally these definitions are incoherently used within philosophy. The approach in this book, thus, is not to hope to come up with the one correct definition, but with a useful one that could be used by many, even if they should use the terms slightly differently.

2.1 A CONCEPTUAL BUFFET
AND A SERVING SUGGESTION

In this chapter, we will discuss the following terminology and recommend settling on some of the definitions currently available. As we have seen in the previous chapter, this step becomes a necessity in order to avoid conceptually confused disagreements about the normative demands and consequences of a machine being "merely interpretable", "explainable but not auditable", "explicable", etc. This chapter includes thus a variety of terms that have only some minor overlap and do not represent competing descriptive accounts of the same phenomenon but rather mark different normative interests within the debate.

Earlier in the explainability debate, authors like Lipton (2018), Doshi-Velez and Kim (2017) have been pessimistic about the possibility of having strict definitions of the issues due to a lack of conceptual clarity. While Lipton bases this diagnosis on a conceptual argument, i.e., that interpretability and explainability cannot be well-defined and thus must leave things with a crucial lack of clarity, it appears that this is merely a problem well-known in other interdisciplinary contexts. Different semantic traditions have produced homonyms of different kinds, often with vastly different background theories, explanatory power, and relevance. Think., for example, about the term "autonomy" in ethical theory, political science, engineering, and sociology. Finding an interdisciplinary definition that does not eventually collapse in homonyms is a challenging task, and the explainability debate is similar in structure: the collection of terms I discuss in this chapter often have their independent different conceptual approaches and terminological choices born from linguistic habits and traditions present in different disciplines, ranging from philosophy to computer and information science, media science, psychology, sociology, and political science.

They differ considerably in the relevance for their field, extent, applicability, and use in these fields and cognitive purposes. Any philosophical work in this which aims to do both translational work between the disciplines and also speak with a certain authority about the matter at hand is best advised in taking a constructive approach. A constructive approach hopes to make the case for a plausible, possible conceptual setup, tries to incorporate many different viewpoints if they can be coherently incorporated, and presents one version of a conceptual setup and terminological choices that work. It does not, however, claim any truth. The appeal of this approach is much smaller than claiming any truths, as we are interested in clarifying and explaining why some normative arguments work the way they work, and why some others may not. I believe the following characterizations allow us to do that, while we should keep the door open for different approaches that might have a different payout in terms of their explainability.

Before we turn towards the normatively relevant concepts, however, we should clarify which technology we are concerned with. The technology these concepts are applied to and argued over are fairly clearly defined: contemporary artificial intelligence, built on machine-learned, or rather by now "deep learned" training methods, fine-tuned with certain goals in mind. As we have seen in the chapter on the metaphor of black boxes, AI is often the paradigmatic case of unexplainable technology. This is striking because until now, our knowledge about the technology we were deploying only grew: we understood better the aerodynamics of airplane wings, improved the efficiency of burning fuel and improved our understanding of the material sciences that permit us to create entirely new kinds of surfaces. And yet, with AI becoming an everyday occurrence in most people's digital lives, we appear to lose knowledge. Not only the day-to-day availability of knowledge, but also, and more worryingly, the principle knowledge of how a technology works.

This "loss of knowledge" is neither new nor especially worrisome in the general sense. With the introduction of commercial aviation by planes, a lot of people began using machinery they would not be able to explain or even be taught to understand. Some of the aerodynamic features were discovered after the fact of creating reliable, useful planes. Many new technologies appear to decrease our knowledge of the technology we are using before it appears to catch up again. There are still new insights about wing designs to be had that explain certain shortcomings of planes (e.g., winglets).

While we will be considering technology that currently is discussed under the label "artificial intelligence", the fact remains that many (digital) technologies currently in use, as well as future ones besides AI (or, worse yet, created *by* AI), are unexplainable to some degree. As we will see, what degree of unexplainability is meant depends in part on the question of when an explanation is sufficient, and what different versions of explanations we aim for. In the next chapter, we will explore the very many concepts suggested in this context and attempt some clarifying work.

2.2   EXPLAINABILITY—SOME DISTINCTIONS

Explainability, in the very rough sense, refers to the fact that something, an event, or a given state or attribute, can be explained. We have several technical definitions from different sources that point towards that general distinction.

Take, for example, Chazette et al.'s (2021) formalized definition, in which they explicitly refer to software systems such as AI (and potentially other ones). In their definition, we find that the explanans of a given explanandum is contextual and addressee-dependent. They claim that explainability is the following:

> A system S is explainable with respect to an aspect X of S relative to an addressee A in context C if and only if there is an entity E (the explainer) who, by giving a corpus of information I (the explanation of X), enables A to understand X of S in C. (Chazette et al., 2021, 200)

Not in contrast, but with a different focus, the International Standards Organization (ISO) defines the difference between explainability and interpretability:

> Interpretability: level of understanding how the underlying (AI) technology works.
>   Explainability: level of understanding how the AI-based system ... came up with a given result. (ISO IEC TR 29119-11:2020(en), 3.1.31 and 3.1.42)

Explainability, here, appears to be more demanding when it comes to granularity than interpretability, as "who the technology works" and "how the system came up with a given result" are two different levels of specificity of the technological workings of a given machine.

These rather rough characterizations do not tell us much about what the explanation for said event or state or system might look like, and what the relationship between those events or states and the ability to explain them consists in. The fact that something is explainable could mean that it is a feature of that thing to be explainable, i.e., it is not fully unexplainable on grounds of its constitution or structure or "essence". It could also mean that it is an ability of the entities attempting to explain the thing, i.e., more akin to a skill or capability. To guide a discussion about explainability, we ought to keep these two perspectives in mind and strictly apart. For this, we may call the fact that the explainability lies within the nature of an event or state "ontological explainability", while the ability to explain such event or state the "epistemic explainability".

2.2.1   Ontological Explainability

Ontological explainability is usually considered to be the core concept of explainability discussions. It means the principal fact that something can or cannot be explained. As we are usually limited epistemically, not ontologically, by our abilities to explain an event or thing, whether unexplainable technologies are "merely epistemically" or "actually ontologically" unexplainable remains an open question. However, some events, at least in human imagination, are by their own nature unexplainable. "God works in mysterious ways" essentially refers to this ontological explainability, and some random events cannot be causally explained but merely be approximated. Albert Einstein's quip of "He (God) does not play dice" is both an admission that Einstein expected nature (and by extension, God) to be explainable beyond a mere approximation of some randomness, as proposed by quantum physicists Max Born and Niels Bohr's interpretation. Whether some artifacts, especially the decision pathways of artificial intelligence, are actually ontologically unexplainable or whether we simply lack the faculty to find the explanations, remains unclear. We may also speculate that artificial general intelligence (AGI) or some quantum computers, if successfully developed, will act in otherwise unexplainable ways. This does not mean that we cannot interpret them (see below), but that they are by their nature eluding an explanation that is gapless.

2.2.2   Epistemic Explainability

Epistemic explainability, in contrast to the ontological one, then, is the idea that we have the necessary epistemic assets and faculties (knowledge, ability to understand, etc.) to explain the event, state, or attribute to the desired cognitive level. Explainable is something if the explainer manages to provide a true explanation at a highly fine-grained level of explanations. If a technology, for example, is epistemically explainable, then this means that we can have justified, true beliefs about the working of the machine. This appears to be the standard level of explainability discussed within philosophy (see, e.g., Wachter et al. [2018], Mittelstadt et al. [2019],

Zednik [2019], Rudin [2019], London [2019]) as well as other areas (e.g., Langer et al. [2019] and others).

These explanatory efforts are also contextualized in our abilities to use available knowledge to explain technologies, as we may be limited in our ability to explain at this stage but with hopes to explain an artifact once we gather more knowledge about the world. The knowledge condition for gapless explanations, then, is problematized in the explanations of the "inner workings" of an opaque algorithm, usually referring to some AI application. These are contemporary paradigmatic cases in which we may seek epistemic explainability with the implicit premise that ontological explainability is present. Or, we might abandon the explicit truth-seeking of explainability efforts in favor of explanations that are as close to actual gapless explanations as possible.

However, under "epistemic explainability" we achieve a gapless explanation from an epistemic perspective if we achieve a desired cognitive level. Since we are bound by the epistemic limits of the human mind, it is often very difficult to discern whether the explanandum in question is ontologically unexplainable or whether we simply lack the ability to do so. Take, for example, Roman concrete (Elsen et al., 2013). The knowledge to reproduce this technology has been lost for centuries and only recently has been figured out again. This technology was, considering the widespread use and success of its materiality, not unexplainable: Roman engineers knew very precisely what kind of ingredients, to which amount, they needed to create concrete that is virtually everlasting (or at least considerably superior to any other material at hand at scale). Yet, for the past couple thousand years, the knowledge to explain and recreate this concrete was lost. It was epistemically unexplainable, not because we could not find the right person who knew how to explain Roman concrete (that we can call "pragmatic unexplainability", see below), but because no person present was able to figure out what had been figured out before (it was potentially even epistemically unexplainable at Roman times, as the engineers could not explain why these ingredients would produce such a lasting material).

These processes constantly appear throughout history, because much of the knowledge required for the explanation of how something is done is, after all, "know-how", and not merely "know-that", which can be recorded easily in books. From our ability to build the pyramids of Gizah, to Stradivari's inimitable violins, to the first computer programs to which the necessary understanding of early coding languages and architectures is lost (as several once popular programming languages have become "dead languages", such as ALGOL), we can gather many examples of lost knowledge that was able to explain certain artifacts to its full, ontological, and epistemic extend (see Olshin, 2019).

2.2.3   Practical Explainability

The level of epistemic explainability is often accompanied by practical considerations since explanations are most often (especially in the context of artifacts) connected to practical purposes: understanding a technology, our own or some other entity's, is rarely ever an end in itself. We would not seek an explanation for an alien propulsion system or language because we want to merely understand its curious nature. We would also seek answers of practical nature: How can it be used? What are the risks associated with this functioning? While explaining the dynamics of the smallest particles in the universe, or the physics at an event horizon at a peculiar constellation in the universe thousands of lightyears away, or the digestive system of Stegosaurus are largely driven by purely cognitive desires, *artifacts* are usually *dealt with* in a manner of practical purposes. Their explanations, then, also carry practical purposes. Practical explainability captures those discussions and usually determines the granularity that an explanation has to provide to satisfy a purpose: depending on when we consider a technology explained in our practical purposes, the ethical demands of explainability are satisfied. This means that for the ethical discussion (Chapter 4) and the applied cases (Chapter 5), we are often dealing with practical demands of the ability to explain the workings of a machine.

2.2.4   Pragmatic Explainability

Next to the practical considerations we should consider that there are pragmatic elements to the explainability of a technology. As mentioned above, it might be true that a given technology is ontologically explainable, i.e., that its principles can be explained step by step without gaps, without it being epistemically explainable, as we simply lack the capability to understand those steps. It might also be that a given technology is epistemically explainable, i.e., that the principles are known and well-documented, that the processes do not exceed human capabilities, but that any such person versed with the technology (i.e., an expert) is unavailable. This distinction goes to demonstrate that explanations, especially of technology, when translated into real-life settings, are constrained by the explanation context in which they occur (or are intended to occur). This will inevitably include a lot of technologies that are in any other notion considered explainable: most technologies in a normal office space are pragmatically unexplainable if the IT person is on vacation, even doctors' offices will have a variety of unexplainable technologies in them and in use: it is probably not too much to speculate that a general practitioner will not be able to explain how exactly a blood test works in detecting all sorts of elements.

2.2.5   Situational Explainability

Lastly, we want to consider situational unexplainability. Situational explainability describes cases in which some immediate, situational reasons may preclude someone from explaining the technology in question, even when all other explainability conditions are met (think, for example, the lack of time or explanatory skill, non-disclosure agreements or other coercive measures, etc.). Considering situational explainability is, in a way, the completion of pragmatic explainability, but stresses that some of the pragmatic reasons we may encounter are not of the structural kind, i.e., the unavailability of health experts due to inequalities of healthcare systems, but immediate ones like the lack of time.

This distinction also stresses that for any applied cases, we ought to be aware that the ability to situationally explain a technology, so to speak "the last mile of explainability", should be considered a necessary condition for labeling something an explainable technology.

It is thus key to a debate to clarify the level of contextually engendered purposes that are discussed to determine the normative relevance of explainability. Most philosophers appear to remain on questions of ontological or epistemic explainability while venturing on occasion into practical or even pragmatic considerations. On the other hand, especially in resource-deprived fields like medical diagnostics and emergency interventions, pragmatic and especially situational explainability are often hard to come by but discussed as a different concern entirely. Being aware of the practical side of an explainability debate might help alleviate some of the preconceived notions about what the findings on more abstract levels might bring. The very fact that a diagnostic tool should, in fact, be made more epistemically explainable than it is currently (see 5.1) might not lead to any changes, let alone improvements, in the treatment sector if situational explainability is not given. If anything, it might hinder medical treatment if based on epistemic explainability a right or claim to explanations is derived that cannot be met pragmatically or situationally. Informed consent, for example, might require situational explainability. If taken seriously, we then might face the fact that some procedures are not being performed out of concern that there is nobody available who can explain the procedure to the patient to the degree required by morality (or law).

Then, in the contemporary debate about the explainability of AI, its technical sub-discipline of xAI, and associated practical undertakings, what is meant, or should be meant, when discussing whether something is explainable or not? As we have seen with the unhelpful metaphor of "black box AI", the problem of explainability emerges in demanding fine-grained explanations for single instantiations of behavior rather than patterns of such behavior. An explainable machine, in this sense, requires the explaining description of the inner workings, the causal processes, and their principles.

2.2.6   Explicability

While the term explainability has taken hold in the debate surrounding opaque technologies, "explicability" was introduced as a normatively loaded notion that aimed at connecting explainability, interpretability, and accountability (Floridi & Cowls, 2019; Herzog, 2022). It was proposed as a concept that includes the descriptive, epistemic challenges of opaque technology (explainability) as well as our practical requirements of understanding (interpretability), and incorporates our responsibility practices to ensure effective regulation (accountability). It was thus aimed at those working on governing AI in different capacities, either as lawmakers, policy advisors, or lobbyists. Explicability is still being used for these purposes, even though the discourse has since reverted to the philosophical challenges that are fundamental to explicability discussions. This synthesized concept is not appropriate for precise philosophical investigations, as its elements are controversially discussed in their own right. Thus, while there have been notable debates on the policy level about explicability of AI, we will prefer to talk about the challenges of explainability even in ethical debates.

2.2.7   Relative Explainability

In Kempt et al. (2022a, 2022b), we develop an account of "relative" explainability which aims at distinguishing between the explainability in context, i.e., relative to practical requirements of a given use-case or even general normative contexts (see, e.g., different norms of depths of explanation for bureaucratic decisions in different societies), and explainability without such context (which we call "absolute explainability"). As we have pointed out above, there is a productive distinction between practical purposes (practical explainability and some subsets [pragmatic, situational]) and more theoretical purposes (ontological and epistemic explainability), outlining the different reasons why one may pursue explainability in the first place.

The distinction between relative and absolute is especially helpful in determining the depth of explanation required when similar human behavior is available and well-established normatively, i.e., in medical decision-making. As we pursue a variety of practical purposes, pointing towards established explanation requirements is a helpful way of deciding whether the establishment of a double standard for AI-explanations versus explanations of machines is justified.

2.3   INTERPRETABILITY

It is important to be aware that interpretability often is used similarly to explainability. As a matter of fact, in several disciplines, interpretability counts merely as a subsection of explainability or denotes exactly what we consider epistemic explainability (see, e.g., Leichtmann et al. [2023] who speak of "explainability" but only discuss interpretability-means). In the former case, explainability is understood as the general concept that defines the task of being able to explain a technology, while interpretability is used as the concept that is then subdivided into different kinds of interpretable models and strategies.

In the latter case, interpretability covers what we consider epistemic explainability, i.e., it delineates the conditions under which we can explain and understand otherwise highly complex machines (Erasmus et al., 2021).

To avoid confusion when talking about interpretability, we thus ought to be clear with what we mean. The understanding I use here underlines the etymology of the term, especially when contrasted with explainability.

While it is true that in any event "explanations" are going to play a key role, not everything we do when reducing opacity should be called "explainability". Different explanations and kinds of explanations will factor in the success of making some technology explained.

However, interpretations, and thus "interpretability", require different epistemic labor from the human side. Interpreting a thing in the world means that the interpreter invests some of their assumptions about that very thing, and/or their background assumptions about the world. Interpretations thus are more about the perspective of the interpreter and their correct theory about the thing in the world and the world around it. Interpretations thus can be explanations, but they are not necessarily connected to the explanandum itself but about its perceived relationship to the world, and the assumptions of those offering the interpretation.

One can easily interpret an event wrongly because one's background assumptions are wrong, or one's theory about the event. Take, for example, my attempt at deciphering a cryptic-looking message from my boss, which contains some errors and seemingly some gaps in thought. I think I know my boss well enough to fill in some of the gaps and correct the typos. One correction, however, is ambiguous and could lead to two different work tasks. Based on my background knowledge of our previous conversation, I am well aware that only one of those tasks is a sensitive task to give to me, and thus I know what task was given to me.

In struggles with the explainability of technology, we often seek explanations that do not require a certain background perspective and interpretation of contexts. The idea of interpreting, rather than "explaining" (in the explainability-sense) technology, lies in the fact that epistemically speaking, we will not be able to provide full explanations and thus are required to interpret decisional pathways of machines. Recall the definition from the ISO: it is characterized as a "level of understanding how the underlying technology works": the level is not about anything machine-specific or internal, but an appreciation of the general principles at work in a machine.

While this sounds like a philosophical position of admitting epistemic "defeat", it is better understood as the description of efforts undertaken at making an unintelligible decision pathway more intelligible (hence, the conceptual confusion emerging between philosophy and computer science): interpretative models of a given algorithm are themselves far less complex than the algorithm they "interpret", and yet they are still explainable in the epistemic sense (see for a discussion on this in Chapter 3, especially with LIME and other techniques).

However, there are practical dimensions to interpretability as well, especially when considering the actual technology in use. In interpretability, not the decisional architecture or decisional pathways available are the subject of interpretation, but the actual decision the algorithm arrives at. Interpreting an algorithm means interpreting the output based on the understanding of the principles of that algorithm, the training data, and the input.

Interpretability, thus, is largely about interpreting a behavior or decision pathway. As most machines do not operate in any "behavioral" sense, they also cannot be interpreted: take, for example, a light bulb that, for some reason, manages to exceed any prediction about its efficiency, without any engineer or researcher being able to explain why. This light bulb may be unexplainable, but it certainly is not uninterpretable (it is, rather, neither interpretable nor uninterpretable: interpretability does not apply to a light bulb).

In human–machine complexes, i.e., coordinative or cooperative instances in which humans and machines make decisions that influence each other, the ability to "read into" (that is: interpret) the reasons for a machine's behavior is key for an expert to work with and rely on that result. The ability to check an algorithm's decision for its "reasoning" is an important means for experts to estimate if their results are trustworthy. Depending on the situation in which this must be checked, it might even affect the expert's ability to take responsibility if the result of human–machine complexes causes harm. Diagnostic AI in clinical contexts, for example, should provide some "reasoning" for doctors to assess the reasonableness of its results to rely on those results. This can be done through some visualization methods such as heat maps for image analysis, for example on an x-ray image. In these heat maps, an algorithm documents the areas that appeared most relevant for its decision-making pathways. It does not specify why these areas were relevant, but they allow for an expert to appreciate the decision-making more than without these maps: a glimpse into the pattern search, the weighing of data against others based on the training, etc.

Interpretability, thus, is not merely about explaining the machine's decision-making differently, but also enabling those who work with these machines to do their work and take responsibility for the consequences of those human–machine complexes.

2.4   OPAQUENESS/OPACITY

We have mentioned several times "opacity" as a term in explainability and interpretability. In opposite to these two, opaqueness is best understood to define the degree of unknown elements of a machine. It is thus closely connected to epistemic explainability and interpretability, as they delineate the conditions and limits of our knowledge, but rather than describing our abilities, opacity describes the state of the matter. Burrell (2016) distinguishes between three different kinds of opacity: (1) one that is caused by secrecy (i.e., trade secrets, patents, etc.), (2) one that is caused by the illiteracy of the users of the technology (i.e., what we call "pragmatically unexplainable"), and (3) the epistemic unexplainable dimensions. While it is sometimes contrasted with transparency (as an antonym), we should understand opacity to describe the state of the unknown elements in a machine, in opposite to transparency, which delineates the normatively problematic lack of knowledge (see section below for details). However, at this stage, we can understand an "opaque system" to be one that is simply not known in further detail, while an "intransparent system" is one lacking details but should not.

There is an individuation challenge in these degrees of unknowability, as the ability to explain or interpret a machine's behavior may require the dissection of specific parts of that machine: a robot, for example, that consists of both mechanical parts (i.e., sensors, actuators, a metal skeleton, etc., i.e., the "hardware") and several complex algorithms governing its behavior (i.e., the "software") might have different degrees of opaqueness in it. Some of the algorithms might be simple expert systems that are well understood while some others are machine-learned algorithms. Equally, some sensor technologies might rely on yet poorly understood processes and materials, while the very basic nuts-and-bolt technology of the exoskeleton is the easiest to explain.

Usually the term "opacity" is thus only used on specified AI-systems that by virtue of their method are highly complex and often impossible to explain (see Chapter 3). This, however, requires a specification of what exactly the opaqueness in a system is, which is often not done (but rather simply applied to the system as a whole).

2.5   TRANSPARENCY

While the term "transparent" is usually used as an antonym to "opaque", in questions of explainable technologies these terms do not function as antonyms to each other. Transparency is a less contentious term in the sense that there is a broader consensus on what it should denote. The question, in contrast to the previous terms, is what is to be made transparent.

The decision-making structure would be a first guess here, but then the question would remain what the difference between transparency and explainability and interpretability would be, which are both in different forms denoting the effort to elucidate the "inner workings" of an algorithm, i.e., to reduce its opacity. Transparency aims at the context of the material construction of the machine—i.e., the means with which such an algorithm was constructed. That translates to demands for the transparency of *sources* of data, their procedural conditions, and potentially any demand made of the producers of data sets, algorithms, and AI (Günther & Kasirzadeh, 2022). This is comparable to the supply chain and manufacturing of certain industrial goods, in which not the inner workings of the industrial product are meant with it being "transparent", but rather that its production process was transparent that it allowed for the control of consumers and democratic institutions.

Two examples currently discussed in the context of transparency of AI are worth exploring in this context, illustrating why transparency, despite its mere loose connection to the other notions, is still a normatively important concept for the explainability debate.

2.5.1   Annotating and the Cost of Human Labor in AI

The conditions of human workers in annotation and human feedback for reinforcement learning, for example, would also fit in this category. As is estimated that thousands of workers work on annotating data sets that are used to train algorithms on (see, e.g., Le Ludec et al., 2023) and that this number will grow due to emerging competition, improved understanding of the efficiency of data sets, etc. In this vein, transparency about the "human cost" and "human labor" needed to produce AI might help contextualize (as a kind of explanation) the technology. Their addition to the transparency requirements could help clarify the nature of this technology.

2.5.2   *Green Sustainability of AI*

As a growing field of study, the environmental impact of AI has yielded some sobering results as the energy and water consumption of the data centers required often has considerable harmful effects on the local environment and the global climate concerns alike (at least until electricity production is fully decarbonized). The desirability of some AI-applications may decrease considerably once the environmental impact is known, and should thus be part of a fully transparent algorithm.

Transparency, thus, is a measure independent of the specific technology, but rather useful for an emerging technology with high externalities. Acknowledging the costs of AI might improve our understanding of how AI works in a socio-economic and ecological context.

2.6   Post-hoc Rationalizability

One of the attempts at explaining the processes in otherwise opaque algorithms is by rationalizing the output of a given process based on the limited knowledge of its construction and the input that produced the output. As this can only be done after the output is known, the explanation will necessarily be less useful than one that has some predictive elements.

This is usually considered a subset of interpretability measures (rather than explainability), as post-hoc explanations attempt to reconstruct decision pathways without the explicit knowledge of the deciding entity (hence the "rationalization"). This requires a certain interpretative stance to be taken towards the rationalized entity, i.e., the technology. As we have seen with the intentional stance (Sect. 1.4) by Dennett, we decide to presuppose some (mental) activities from which we can delineate predictive or at least explanatory powers. It is thus inherently interpretative, as we do not know, nor do we have to know, anything about the deciding entity. In the case of rationalization, the perspective taken is one of AI as a rational decider, from which we can infer certain decisions based on the output produced.

As it is rule-based, i.e., rational, the explanations given are not "propter hoc", i.e., fitted towards a preconceived explanatory goal. The idea of strictly rational post-hoc explanations is that these explanations only provide explanatory insight if their argumentation is limited to rational principles. This does not mean that these principles are those of rational decision-makers (i.e., the intentional stance), but that the means to interpret an AI's decision-making are limited to previously agreed upon rational principles (for an overview, see Gurrapu et al., 2023).

Illustrative examples come from users asking trick questions to general-purpose chatbots. While one would expect them to either fall for the trick question or to avoid it, pointing out that it was indeed a trick question, the answers are even more illuminating when chatbots claim that they figured out the trick question, even when the trick of the trick question was that it was merely formulated like a common trick question. We see that the chatbot still insists on having figured out the trick of the trick question while, logically, there was no trick. What appears to be an almost human flaw, i.e., impatience and some mild arrogance of already knowing the answer even when the wording is slightly changed, post-hoc rationalization helps to clarify that the lack of common sense and the training on the standard formulation of trick questions has warped the chatbot's answer into being false.

Thus, post-hoc rationalizations are those explanations that provide a rule-based interpretation of the behavior and decisional pathways of autonomous technologies that fit the outcome based on the limited knowledge about the technology as well as the input.

2.7   AUDITABILITY

The ability to audit an AI's decision-making pathway is often considered the legal requirement of explainability standards. This is to be understood in opposition to contestability (Sect. 2.8), in which the ability of decision subjects to question the decision pathways of an algorithm is the point of concern. Auditability, however, refers to the ability to exert collective, democratic control (Toader, 2020).

Auditability touches upon the larger decisional architecture in the algorithm, as well as its resilience towards adversarial attacks, the control for and reduction of biases, and reliability in decision-making. With many algorithms making decisions that have either immediate, unsupervised consequences (i.e., unchecked decisions about loan or parole applications) for their decision subjects, or that can compound their harm (i.e., content-recommender algorithms), the ability to control the decision-making processes of such algorithms for unbiased and fair decisions is a key part of auditability.

It is not surprising, thus, that auditability is usually discussed in the context of AI safety considerations, as the ability to audit an algorithm gives those developing safety standards the ability to test them against the currently available technology. AI safety, in this sense, is not asking whether an AI ought to be explainable, and if the AI should be used, but rather what technological means ensure the liable development of an AI for broader use. How specific AI needs to be explainable to satisfy auditability requirements is to be determined by those determining the standards, i.e., ultimately the legislative body.

This is because auditability is also connected to questions of transparency, e.g., the willingness of companies to offer their code for legal revue in cases of damages caused by the autonomous technology (i.e., to clarify liability questions after legally relevant errors have occurred, and to resolve criminal or tort law questions). Auditability, thus, ensures the legal liability of AI companies and their products as there can be specific requirements formulated that have to be fulfilled to be able to be certified (see Sect. 2.9).

2.8   CONTESTABILITY

Contestability is being put forward as a means to relegate the source of the normative weight of an AI's decision (or decision suggestion) to those affected by the decisions. These "decision subjects" ought to have, in this perspective, the ability to contest the decision of an AI (or otherwise opaque decision-making system). The ability to contest a decision, however, requires the capability of the system to be explained to a degree that makes any challenges to the decision decidable. Contestability establishes a normative claim of explainability of those who will be subjected to the decision against the system: it is not about merely understanding how the decision came about (as could be achieved in post-hoc-rationalization), but to appreciate the decision pathways in a way that enables a potential reason-based rejection of such decision.

This position is largely motivated by processes in medical decision-making (e.g., Ploug & Holm, 2020), but might apply to other areas of opaque decision-making processes as well. Take, for example, opaque decision structures in granting an applicant a loan or a job interview, or a prisoner parole. These decision-making procedures have been beset with biases and unjustifiable heuristics, and insisting on algorithmic decisions to be contestable merely ensures that we do not re-introduce these biases and heuristics back into our decision-making procedures under the guise of objectivity and impartiality. According to Ploug and Holm (2020), there are four criteria to be met for their "effective contestability" that would satisfy explainability requirements: (a) the system's use of data, (b) potential biases, (c) performance, and (d) the amount of shared labor between medical personnel and the machine. However, they stress that these criteria are domain-specific to questions of medical diagnostics and AI.

2.9   CERTIFIABILITY

Certifiability demands that a technology or a technologically influenced decision-making process ought to be assessable based on a public evaluation of the accuracy, reliability, and associated risks of its use (see Tutt with an early proposal for an "FDA for algorithms", Tutt [2016]). In opposition to contestability, certifiability does not provide decision subjects of an unexplainable technology (or processes that embed these technologies) with a claim to reject that technology's decision merely based on its lack of explainability (though it may be based on other concerns).

Introduced as a pragmatic concession in light of ubiquitous unexplainable processes in the practical and pragmatic explainability sense (cf. Sect. 2.2), certifiability shifts the perspective from the rights and claims of decision subjects towards a reliable method to deal with their remaining risks (Landgrebe, 2022). This key insight here lies in the fact that for many decision-making contexts that involve a certain amount of modern digital technology, the decisions will be practically and pragmatically unexplainable. In these many cases, however, we accept that there are stable trust networks of those explaining an almost unexplainable decision-making process (i.e., to the degree laid out when talking about epistemic explainability) and those affected by those decision-making processes by way of the former still taking responsibility for negative outcomes affecting the latter (as we do with heuristics and other not epistemically explainable decision-making procedures). This allows for the normatively seamless integration of many modern technologies into established interactive contexts. However, concentrating on these potentially negative outcomes refocuses the debate about the appropriate target of normative consideration from a duty to explain to a comprehensive risk-assessment.

Managing these risks, however, cannot reasonably be relegated to the decision subjects, who may be overwhelmed by the amount of information necessary to appreciate the risks and who may draw the wrong conclusions about the risks presented to them. Additionally, as many technologies' intended use implies their widespread dissemination, a collective response to the risks these technologies pose is required, in opposition to an individual's decision.

Certification is an option to determine acceptable risks for those using and being subjected to certain technologies by way of democratic discourse. This means that the explainability of a technology is only relevant insofar as it allows the public, and in focus the lawmakers, to determine the reliability of a technology and the risks associated with its (un-)reliability. It then is up to the public to determine the acceptability of certain risks in light of the utility of the technology in society at large. This makes explainability only an instrumental good for other purposes more relevant to the use of said technology.

Clearly, different use cases for unexplainable technologies, such as medical decision-making or autonomous weapons systems, carry different requirements for acceptable risks and, thus, their certifiability. We can also see that the engineering quality and thus the risks associated with similar technologies differ from instance to instance. Two different diagnostic AI tools, for example, may exhibit considerably different error types and error rates, illustrating the need for a social consensus on what should be included, measured, and limited in a certification (however, this also demonstrates that such a determination can look very different between different legal entities and their collective moral expectations and tradition).

One major challenge to certifiability is the precise definition of the purposes and the performance of a machine. This means that for those applications of AI, which have little to no specific purpose but rather represent the state-of-the-art, i.e., foundation models, the sufficiently precise definition of their purpose may be impossible, rendering them potentially "uncertifiable". These models are, however, instrumental in creating applications that may be desirable, which would render the "uncertfiability" a problematic practical consequence.

Thus, certifiability cannot be a means to determine an industrial standard within engineering departments but functions rather as a moral demand that the risk-assessment of unexplainable technologies are necessary requirement for the justifiable use of these applications in contact with users and consumers. As Brajovic et al. (2023) suggest, this is a merging between engineering departments and political institutions.

2.10   Interim Conclusion

We have seen that the very many accounts, approaches, depths, kinds, and normative demands for explanations have created a massive discourse about how to deal with technology that is partially or fully unexplainable according to these requirements.

However, what has become clear here as well is that the different kinds of algorithms and their use trigger different kinds of explanations. Not because the normative contexts are differing, but because of the nature of the decisions they are making: a sorting-and-recommender algorithm has fundamentally different capacities to be explained than, say, a general-purpose chatbot, both due to complexity and the institutionalized use of these algorithms. Thus, the variety of approaches can be explained by the different attempts to deal with the issue of using unexplainable technology for practical purposes.

As we will see later on, these kinds of explanations required for the justified use of AI are dependent on an even larger variety of different conditions. Not only is the kind of decision relevant, but it also depends on whether such decisions are used to replace human decisions, whether this has an effect on a societal level, and so on. Merely listing the different accounts will not get us an insight into explaining unexplainable technology at large, mostly because that will lead to epistemic explainability requirements which often are a fool's errand.

References

Brajovic, D., Renner, N., Goebels, V. P., Wagner, P., Fresz, B., Biller, M., Klaeb, M., Kutz, J., Neuhüttler, J., & Huber, M. F. (2023). *Model reporting for certifiable ai: A proposal from merging eu regulation into ai development.* arXiv preprint arXiv:2307.11525

Burrell, J. (2016). How the Machine 'thinks': Understanding opacity in machine learning algorithms. *Big Data & Society, 3*(1), 1–12.

Chazette, L., Brunotte, W., & Speith, T. (2021). Exploring explainability: A definition, a model, and a knowledge catalogue. In *IEEE 29th International Requirements Engineering Conference (RE)* (pp. 197–208). IEEE. https://doi.org/10.1109/RE51729.2021.00025

Doshi-Velez, F., & Kim, B. (2017). Towards a rigorous science of interpretable machine learning. *arXiv* preprint https://arXiv:1702.08608.

Elsen, J., Cizer, O., & Snellings, R. (2013). Lessons from a lost technology: The secrets of Roman concrete. *American Mineralogist, 98*(11–12), 1917–1918.

Erasmus, A., Brunet, T. D. P., & Fisher, E. (2021). What is interpretability? *Philosophy & Technology, 34*, 833–862.

Floridi, L., & Cowls, J. (2019). A unified framework of five principles for AI in society. *Harvard Data Science Review, 1*(1). https://doi.org/10.1162/996 08f92.8cd550d1

Günther, M., & Kasirzadeh, A. (2022). Algorithmic and human decision making: For a double standard of transparency. *In AI & Society, 37*, 375–381.

Gurrapu, S., Kulkarni, A., Huang, L., Lourentzou, I., & Batarseh, F. A. (2023). Rationalization for explainable NLP: A survey. *Frontiers in Artificial Intelligence, 6*.

Herzog, C. (2022). On the risk of confusing interpretability with explicability. *AI Ethics, 2*, 219–225. https://doi.org/10.1007/s43681-021-00121-9

Kempt, H., Freyer, N., & Nagel, S. K. (2022a). Justice and the normative standards of explainability in healthcare. *Philosophy and Technology, 35*(100). https://doi.org/10.1007/s13347-022-00598-0

Kempt, H., Heilinger, J. C., & Nagel, S. K. (2022b). Relative explainability and double standards in medical decision-making. *Ethics and Information Technology, 24*(20). https://doi.org/10.1007/s10676-022-09646-x

Landgrebe, J. (2022). Certifiable AI. *Applied Science, 12*(3), 1050. https://doi.org/10.3390/app12031050

Langer, M., Oster, D., Speith, T., Hermanns, H., Kästner, L., Schmidt, E., Sesing, A., & Baum, K. (2021). What do we want from explainable artificial intelligence (XAI)?—A stakeholder perspective on XAI and a conceptual model guiding interdisciplinary XAI Research. *Artificial Intelligence, 296*. https://doi.org/10.1016/j.artint.2021.103473

Leichtmann, B., Humer, C., Hinterreiter, A., Streit, M., & Mara, M. (2023). Effects of explainable artificial intelligence on trust and human behavior in a high-risk decision task. *Computer in Human Behavior, 39*, 107539. https://doi.org/10.1016/j.chb.2022.107539

Le Ludec, C., Cornet, M., & Casilli, A. A. (2023). The problem with annotation. Human labour and outsourcing between France and Madagascar. *Big Data & Society, 10*(2). https://doi.org/10.1177/20539517231188723

Lipton, Z. C. (2018). The mythos of model interpretability: In machine learning, the concept of interpretability is both important and slippery. *Queue, 16*(3), 31–57. https://dl.acm.org/doi/pdf/10.1145/3236386.3241340

London, A. J. (2019). Artificial intelligence and black-box medical decisions: Accuracy versus explainability. *Hastings Center Report, 49*(1), 15–21. https://doi.org/10.1002/hast.973

Mittelstadt, B. D., Russel, C., & Wachter, S. (2019). Explaining explanations in AI. *Proceedings of the conference on fairness, accountability, and transparency* (pp. 279–288). https://doi.org/10.1145/3287560.3287574

Olshin, B. (2019). *Lost knowledge: The concept of vanished technologies and other human histories.* (Vol. 16). Brill.

Ploug, T., & Holm, S. (2020). The four dimensions of contestable AI diagnostics—A patient-centric approach to explainable AI. *Artificial Intelligence in Medicine, 107*, 101901. https://doi.org/10.1016/j.artmed.2020.101901

Rudin, C. (2019). Stop explaining black box machine learning models for high stakes decisions and use interpretable models instead. *Nature Machine Intelligence, 1*(5), 206–215.

Toader, A. (2019, November 11). *Auditability of AI systems—Brake or acceleration to innovation?* https://doi.org/10.2139/ssrn.3526222

Tutt, A. (2016). An FDA for algorithms. *Administrative Law Review, 69*(1), 83–123.

Wachter, S., Mittelstadt, B., & Russel, C. (2018). Counterfactual explanations without opening the black box: Automated decisions and the GDPR. *Harvard Journal of Law & Technology, 31*(2), 842–861. https://doi.org/10.2139/ssrn.3063289

Zednik, C. (2019). *Solving the black box problem: A normative framework for explainable artificial intelligence.* arXiv. https://arxiv.org/abs/1903.04361

CHAPTER 3

Epistemological Conditions

Abstract In this chapter, we explore under what conditions explanations about technology amount to knowledge, and how this knowledge can be used further. In first determining that the level of granularity is already a purpose-driven one, we establish that there is no explainability without predetermined epistemic of practical interests. In pointing towards the different epistemic statuses of explanations and interpretations, we determine that interpretations do not provide knowledge-based explanations, but still produce some insight that are useful for practical purposes. Lastly, we turn to the question whether unexplainable technology can contribute to our scientific research, considering the results cannot be explained as other scientific knowledge is explained.

Keywords Explainability · Interpretability · Granularity · Knowledge generation

We have seen that we require occasionally vastly different kinds of explanations to arrive at a satisfying answer about the decision pathways of an autonomous machine and its algorithms. These are usually motivated by the equally different kinds of practical purposes pursued in demanding or finding an explanation.

© The Author(s), under exclusive license to Springer Nature Switzerland AG 2024
H. Kempt, *(Un)explainable Technology*,
https://doi.org/10.1007/978-3-031-68098-4_3

For most discussions about the explainability of technology, epistemic explainability seems to be the target level of explanation. While we discuss the relevance of epistemic explainability for normative considerations here and in the next chapter (Chapter 4), delineating the epistemological conditions of explainable technology is a philosophical challenge of its own. This is because some of the practical explainability requirements hinge on the answers given in the determination of the limits of epistemic explainability.

In this chapter, we will first examine the question of which discipline is called upon to explain unexplainable technology, to then specify the level of granularity in which these explanations should be kept to meet our epistemic purposes. Afterward, we assess some interpretative models and then conclude that the knowledge, present in explanations opposite to interpretations, is the challenge for unexplainable technology. We then discuss how the expectation of knowledge-based explanations, however, can create some subtle normative requirements. Lastly, we discuss how these explanations factor in using unexplainable technology for primarily cognitive purposes.

3.1   Unexplainable Technology: Whose Jurisdiction?

What can philosophers even say on the matter of unexplainable technology, if that matter is not somewhat related to the ethics of developing and using such technology? Tomsett et al. (2018) distinguish between six different types of stakeholders that pose epistemic demands, which explains the vast difference in conceptual and terminological setups when approaching unexplainable technology. Their types include developers, examiners, operators, executors, data subjects, and decision subjects. As Kasirzadeh points out (Kasirzadeh, 2021), this view, and the models serving this view, fall short of providing the answers we may seek as philosophers or social scientists, and presuppose limited stakeholder-views of those immediately involved, while we might have society-wide standards and values that ought to be incorporated. However, this does not clarify what epistemic role philosophers can play in gaining insight into explainability.

The imperative to improve our understanding of said technology is uncontroversial and lies firmly with the computer science departments and AI companies around the world, while the implementational questions are relegated to other stakeholders—even if we expand the list to include ethicists.

It is best for computer scientists to conceptualize the question of what philosophers can contribute from the open question emerging when asking the opposite side: What cannot be explained, and why? This provides an insight into the limits of explanations currently available, and, thus, some of their qualities (or lack of qualities). However, answering this question with the precision needed is not as easy as it initially seems: Are our explanations insufficient because we do not yet have a working theory of the thing in question, or can there even be an explanation that reaches the levels we demand? If the latter was the case and computer scientists would claim something being "unexplainable" as a property of the thing, then we would face what we dubbed above "ontological unexplainability". Thus, to achieve explanations that serve the often implicit or presupposed epistemic purposes, we should make these explicit. In this sense, philosophers' analyses of the different kinds of explanations needed for different epistemic and practical purposes are best understood as an assistance to the effort. Philosophers could not possibly hope to understand all the available model-agnostic (or even model-specific) interpretive approaches and the specifics of machine-learned AI, in general, to comment on whether the explanations provided in those accounts are, in fact, knowledge-producing or not.

Therefore, the key question that philosophers can contribute to the pursuit of creating explainable technology, especially the large field of explainable artificial intelligence (xAI), is the determination of the epistemically adequate levels of explanation provided as a result of this effort. In this book, this has been undertaken in a first step by the reconstruction and creation of a common ground through shared vocabulary and a coherent conceptual approach, even if this approach does not mirror precisely the debate as it is being conducted (from Chapter 2).

The explication of the epistemic purposes, then, is the next step for the philosophy of science in the struggle for explainable technologies. I argue in this next step that the concept of granularity is an illustrative way to explicate these epistemic purposes, and especially point towards the fact that these purposes, just because they are epistemic and cognitive in nature, do not require the smallest level of possible explanation.

3.2   A Question of Granularity

Explanations come in many different forms, and most of them serve a purpose that is largely practical. However, as we have seen in Sect. 1.3, a "good" explanation is not only one that serves specific practical purposes, but rather provides some insight that is both cognitively appealing as well as enabling some practical purposes. The better the theory that explains a given phenomenon, the more can this phenomenon be expected, predicted, or potentially even controlled.

In comparing traditions of explanations depths in other disciplines, we quickly notice that there are some inner differentiations according to the explanatory frame of the discipline and its presupposed cognitive purposes. Take, for example, the explanation of a chemical experiment and the processes behind it. In most contexts of chemistry, the features of particular molecules, the forces present and applied, the purity of the resources, etc. are featured in the explanation for a certain reaction. Those same contexts also reject explanations that go below that level, as this would be more akin to doing particle physics (i.e., describing a process and its explanations merely in terms of forces). Next, the fact that we still have a very poor understanding of how the causal relationships between atomic particles and the non-causal relationship sub-atomic particles have can be explained (i.e., how the laws of physics scale down to the smallest possible level), the levels of explanation are chosen by the epistemic purposes pursued in the scientific activity. This, however, means that they are fundamentally purpose-driven. This is almost trivial from the perspective of the philosophy of science, as we can only find answers to the questions we are asking, and thus by asking specific questions, we also only find specific answers. And yet this is not trivial in trying to settle on levels of explanations, or as we shall call it here, the *granularity* of explanations, as they require a specific definition of the purposes in question: What is it exactly we want to know, and to what degree of justification do we want to know it for it to be justified? With the determination of those questions, we also set the level of granularity required (and also on the side the other conditions we expect from the explanation besides its granularity, such as causal or normative explanations).

Another example of how these epistemic purposes are often even driven by practical purposes (or at least can have some substantial practical ramifications that can influence the epistemic purposes) is the rather innocent question of "how long is the shoreline of Great Britain?" This is not a political question (i.e., testing whether someone knows the differences between the British Isles, Great Britain, the UK, England, and so on), as so much a mathematical one.

Known as the coastline paradox (first observed by Hugo Steinhaus [1954]), the challenge lies in the length of the basic unit with which one measures the coastline (or any other measurement). Seemingly paradoxically, the more fine-grained one measures (i.e., the smaller the smallest unit of one's measurement), the longer the coastline becomes even though, quite obviously, the coastline remains the same. The lengths of coastlines are thus defined by the purpose of measuring them. This sounds like there is not an "objective" or "true" length of Britain's coastline, which there is of course. There are true statements about the length of the coastline. However, there are very many different ones, and agreeing beforehand which is the basic unit of measurement determines the outcome. Such agreements, however, often turn and thus depend on practical purposes, e.g., the longer a country's coastline, the more fishing rights it will be granted in international waters.

This does not mean that all epistemic purposes are ultimately practical purposes (even though some pragmatists appear to claim that) unless we want to consider "a shared language with which to talk about scientific discoveries" to be a practical rather than an epistemic purpose. There are epistemic purposes that evade a reasonable connection to practical purposes: from explaining the grammar of a dead language with no written texts to the precise interaction of celestial bodies millions of lightyears away, etc.: fundamental or niche research is often struggling to find any particular practical purpose with the knowledge gained. This also applies to some everyday life experiences we might have, in which we have epistemic, cognitive purposes that serve no obvious practical purpose. Think of doomscrolling and other uses of social media, in which we expose ourselves to worldwide news and content with no explicit practical purpose. Some things, it appears, we simply *need to know*.

Granularity is not limited to mere epistemic purposes either: the explanation to the question "Why is the water warm?" can have contextual features that determine the kind and granularity of explanation required: "The water is warm because the water particles are moving faster than before" is as much a "correct" answer as is "the water is warm because the stove is on and the container in which the water is held transferred the heat from the stove to the water", as well as "the water is warm because I am making some tea" (if the person is indeed making tea, of course).

In order to move towards using granularity as a measure to determine the kinds and extends of explanations necessary to consider a technology explained, we need to determine what level of granularity is acceptable. I argue that gaplessness, as introduced in Sect. 1.3 as part of a good explanation, is a rule that provides this kind of determination: If an explanation can, within its own established level of granularity, explain a phenomenon without leaving gaps in the explanation, then it merely remains to those seeking an explanation to determine which level they may need.

Ruling out the switch from one explanatory frame to another guarantees explanations to remain coherent, it also differentiates that the pursuit of explainable artificial intelligence (xAI) can come in different degrees. xAI is not a monolithic block of the same level of explanations, but a research program that has several inner differentiations (cf. Byrne, 2023, 6536–6537).

Being able to provide a gapless explanation on a more fine-grained level thus would constitute progress within this debate: the more fine-grained an explanation can be provided about the decision-making pathway of an AI without resorting to speculative or undetermined parts within this process, the more progress is achieved in explaining AI. This explains how xAI has become both a practical effort, but also its own scientific endeavor, which—like most scientific efforts—can become rather niche. After developing reliable, knowledge-based explanations on a larger level, the level of granularity may become so small that the explanations provided serve only a limited epistemic purpose of specialists (akin to progress in figuring out a rule of past tense in a dead language's grammar), and no immediate or even mediated practical purpose down the line. We might just understand AI better, but have since moved on from that debate as a normative concern. Epistemic explainability here means delineating, but also reassessing the possibility of creating knowledge-based explanations without explanatory gaps.

The issue with the computer science-debate surrounding explainable AI, on the one side, and the philosophical discussion, on the other, is that they seem to pursue different kinds of explanations, different levels of explanations, and different granularity of explanations. For a successful discourse to take place, these have to be settled first (this is also reflected in the challenges of conceptual setups, see Chapter 2). In the following, we are discussing how an interpretative model demonstrates this distinction (and why it may become difficult to achieve epistemic explainability with interpretative models).

3.3   LIME AND OTHER INTERPRETATIVE MODELS

We have seen in the previous conceptual distinctions that interpretability, i.e., the ability to provide a *working theory* about the inner workings of a machine based on the information retrievable, is firmly settled within epistemic conditions about our ability to know. In this vein, the research program of xAI has unfolded partially in pursuit of creating interpretable models, i.e., models that can predict the behavior of a different model while being explainable itself. This means the reduction of complexity in the inner workings of one machine that still manages to predict the output of another, which represents an interpretation, as the prediction mechanism does not use the same inner architecture as the original one.

One of the most successful approaches of this kind is called "LIME", i.e., "local interpretable model-agnostic explanations", which can explain individual outputs of any given unexplainable model while remaining explainable (Ribeiro et al., 2016). The fact that this approach is "model agnostic" by the authors' terms demonstrates, on the one side, that such an approach can work with many different models, and on the other side, that there is little knowledge generated about the inner workings of the target model—these explaining models remain "agnostic", after all. Other models include heat maps (e.g., Tjoa et al., 2021). These show how an image-recognition and sorting algorithm "read" the picture in question by pointing out which areas of the image were especially important for sorting into a specific category. This, similar to lime, may even predict the algorithm's decision by pointing towards statistical methods of similar heat maps. However, it still only gives us "superficial", interpretative insight into what an algorithm did or is about to decide, without telling us more in detail (i.e., more granular) how it decides what it decides.

While their use (and thus their explanatory power) can come with some pitfalls, there is reason to believe that proper use of the growing number of model interpretations can achieve a reliable explanation of the predictions made by the explananda, i.e., the larger, opaque models (Molnar et al., 2022).

There are approaches to causal AI, i.e., a reconstruction of causal inferences within neural networks to determine why a certain decision has been taking the path it took (Zenil, Kiani & Tegner 2023). These approaches aim at providing the full picture of a machine-learned AI's decisional pathway, including how a certain input led down the path through the nodes towards the output. Whether any of these approaches are likely, or even probable, to succeed remains to be seen. The fact that some computer scientists have not given up the attempt at fully, gaplessly, finely granular explaining AI's decision-making pathways is admirable; however, the chances of success of this account for currently deployed large models are doubtful.

Epistemically speaking, we may have not achieved the explanatory depth we desire for full epistemic explainability. We certainly gain insight into some of the workings of the target machine by being able to predict its output to a higher degree; however, what we have found is not a causal explanation for the actual inner workings of said machine, but rather a simpler model that has output that is highly correlated in its output to the one we seek to explain. We did mention above that prediction or predictability is one key dimension of explanations: a good explanation allows not only to post-hoc-rationalize what occurred but also to make predictions about similar future events. In this sense, interpretable models are very useful in improving our understanding of some unexplainable technology.

3.4   KNOWLEDGE IN EXPLANATIONS?

We now ought to turn our attention to the question of whether we generate the right kind of knowledge in these explanations. As we have stated before, explanations are closer to truth claims than interpretations about the same phenomenon: explanations should be correct explanations, usually about the connections between cause and effect. Interpretations, on the other hand, offer an explanation from a perspective under given limited evidence with potentially non-truth-related purposes. That does not mean that interpretations are merely ways of how one subject relates to an event, but that the explanatory power of interpretations lies with something other than its truth (even though interpretations are assertions and thus truth-conditional).

Model interpretation methods like LIME, for example, are usually considered interpretative tools as they provide (or maybe *restore*) some predictive powers back to the technology-use, without providing much about the truths of the decision-making process in question. These approaches enable us to handle these ever-growing models, i.e., to interpret them. What they do not provide is knowledge about how these models do it.

The question, ultimately, is to what degree of granularity can we provide explanations that are still truthful in the sense we inquire? As we have seen, the granularity cannot (and must not) go to an atomic level, since the explanations would be useless and out of touch with our practical understanding of what explanations ought to achieve, especially in the domain of explanations of technological behavior. Nobody would be satisfied if we would have to consult physicists in a self-driving car's decision to turn left on a right-turn-only lane, as we examine the flow of electricity in the moment of the decision. However, it also cannot remain on a too abstract or general level, as those explanations do not have the explanatory power we seek as well (imagine a philosopher merely pointing out the limits and risks of self-driving cars due to their lack of common sense understanding).

A "standardized level of inquiry about singular decisional pathways" appears to be the adequate level of explanations for autonomous technology: that level allows us to work with practical purposes but also should cover our cognitive ones without expanding those into even more basic questions of the nature of computer science, electricity, or so (explaining a decisional pathway of a technology should not require to explain the underlying physical principles).

3.5   The Subtle Normativity of Epistemic Explainability

As we have pointed out several times already, determining the extent to which one can explain a certain technology is largely a descriptive endeavor but also comes with normative ramifications: in elaborating how epistemic justifications for the explanations of technologies work (i.e., how we can be justified in believing to understand precisely what is going on, and what can count as a gapless explanation), we also deliver some of the practical dimensions of explainability from the previous chapter. If we cannot know some explanatory dimensions on a certain level of granularity, but determine that that level is required for a justified use of that technology, then we may face justificatory troubles for using that technology.

This has been a controversy in, for example, medical diagnostics as we demand knowledge-based explanations for the decisions of doctors. In this picture, informed consent can only be based on the knowledge provided to the patient, i.e., we demand epistemic explanations for certain assessments. Interpretability, as discussed above, does not produce this kind of knowledge, even when doctors still claim responsibility for the decisions made.

Knowledge-based explanations, rather than approximations and interpretations, appear to carry some independent normative weight, even if the practical purposes of reliable approximations, interpretations, and agnostic but reliable predictions outweigh the need for knowledge-based explanations.

This is reminiscent of the oracle-intuition from Sect. 1.2, in which we showed that there are justificatory pressures for non-knowledge-based decisions, even when they appear to be most reliable (or even just more reliable than our current knowledge-based approaches), to become more explainable, i.e., knowledge-generating. Additionally, these pressures persist even when there are currently few means to provide such explanations: we are called upon to attempt to provide these explanations.

3.6   INTERPRETATION-BASED SCIENCE?

There is another dimension of creating technology that is unexplainable but can be used in chiefly cognitive tasks, like scientific research, especially when there is virtually no means of validating new AI-driven insights (Durán, 2021): if we cannot explain the decision-making process of an AI that might provide human researchers with key insights into their theories, then how can we hope to produce knowledge in science? The concern here develops not from using technologies on human beings or in processes involving human beings (i.e., largely led by practical purposes), but in processes that are cognitive in nature: knowledge-based explanations are the key for scientific processes, even if scientific *discoveries* or *research hypotheses* can come about through mere trying, educated guessing, or accidents. The scientific reasoning behind those hypotheses, however, requires explanations in line with the scientific method.

This, however, appears to become the case here when relying on AI for doing research, especially in situations where either the cognitive skills of humans, or their ability to intervene and guide the process are hampered (e.g., by doing higher-level maths at speeds that cannot be appreciated by humans, or by doing self-guided scientific experiments in space). This could, in the worst case, relegate our scientific reasoning from causal inferences to statistical inferences and associative explanation, in which we basically merely build interpretable models that provide approximations of the original AI, while leaving the actual scientific reasoning process of the AI in the dark (if one could speak of "scientific reasoning" in this context at all).

As we have stated in Chapter 1, a good explanation is appreciable by more than just the person providing the explanation: the reasons for an explanation, thus, ought to be intersubjective, if not objective in nature. In scientific reasoning, these reasons might become driven and provided by AI, and thus unexplainable to researchers altogether.

This worry, as has been articulated in the philosophy of science (especially powerful in Messeri and Crockett [2024]), might be a bit over-stated considering the highly complex, technology-reliant research that is currently already done. Take, for example, much of the astrophysical data that is being filtered and processed is handled by layers of technology to which there is usually neither the time nor the reason to check for correctness. The process of data gathering, and even independent data analysis, is performed by neural networks that, in this original, granular level of explanation we have settled on, is unexplainable. And yet, the results, and even the astronomers' explanations, appear as resilient to skepticism as before.

The question of research is still posed by human scientists who have a map of the incompleteness of the scientific knowledge, and thus are unable to conduct their research on those areas, or those that—in light of new evidence and theories—have become less explored yet again (see, e.g., Grossmann et al. (2023) for a look on social science research using generative AI, or Wang et al. (2023) for a look on possible scientific improvements across disciplines). Further, the idea that scientists merely produce knowledge-based explanations is somewhat idealizing, as many explanations are motivated by non-epistemic reasons, such as the psychological pressure to publish statistically significant results (via p-hacking or some similar data-dredging methods of manipulating one's research success, see Smith, 2014 for a wide-ranging analysis).

Thus, in an objective sense of scientific research, unexplainable research technology may be an issue for the results it may provide, as we lack the ability to test these results for their truth. However, as we have seen, most science simply does not work like this, and unlikely ever will. The nobleness of the sciences in this objective sense is rarely ever achieved, as scientists interpret and explain at the same time. This does not limit their validity claims, as both their interpretation and explanations are generalizable; however, we should remain skeptical of those that condemn the scientific process if we introduce unexplainable technology: that, what is ultimately explained, can be identified by something unexplainable.

REFERENCES

Byrne, R. M. (2023). Good explanations in explainable artificial intelligence (XAI): Evidence from human explanatory reasoning. In *Proceedings of the thirty-second international joint conference on artificial intelligence. International joint conferences on artificial intelligence organization* (pp. 6536–6544). Macau, SAR China.

Durán, J. M. (2021). Dissecting scientific explanation in AI (sXAI): A case for medicine and healthcare. *Artificial Intelligence, 297*, 103498. https://doi.org/10.1016/j.artint.2021.103498

Grossmann, I., et al. (2023). AI and the transformation of social science research. *Science, 380*, 1108–1110.

Kasirzadeh, A. (2021). *Reasons, values, stakeholders: A philosophical framework for explainable artificial intelligence.* https://arxiv.org/pdf/2103.00752

Messeri, L., & Crockett, M. J. (2024). Artificial intelligence and illusions of understanding in scientific research. *Nature, 627*, 49–58. https://doi.org/10.1038/s41586-024-07146-0

Molnar, C. et al. (2022). General pitfalls of model-agnostic interpretation methods for machine learning models. In A. Holzinger, R. Goebel, R. Fong, T. Moon, K. R. Müller & W. Samek (Eds.), *xxAI—Beyond explainable AI. xxAI 2020.* Lecture Notes in Computer Science(LNCS), vol. 13200. Springer. https://doi.org/10.1007/978-3-031-04083-2_4

Ribeiro, M., Singh, S., & Guestrin, C. (2016). *Why should I trust you? Explaining the predictions of any classifier.* arXiv. https://arxiv.org/pdf/1602.04938.pdf

Smith, G. (2014). *Standard deviations: Flawed assumptions, tortured data, and other ways to lie with statistics.* Overlook Press.

Steinhaus, H. (1954). Length, shape and area. *Colloquium Mathematicum, 3*, 1–13. https://doi.org/10.4064/cm-3-1-1-13

Tjoa, E., Khok, H. J., Chouhan, T., & Cuntai, G. (2021). *Improving deep neural network classification confidence using heatmap-based eXplainable AI.* arXiv preprint arXiv:2201.00009

Tomsett, R., Braines, D., Harborne, D., Preece, A., & Chakraborty, S. (2018). *Interpretable to whom? A role-based model for analyzing interpretable machine learning systems.* arXiv preprint arXiv:1806.07552

Wang, H., et al. (2023). Scientific discovery in the age of artificial intelligence. *Nature, 620*, 47–60.

Zenil, H., Kiani, N. A., & Tegner, J. (2023). *Algorithmic Information Dynamics: A Computational Approach to Causality with Applications to Living Systems.* Cambridge University Press. https://doi.org/10.1017/9781108596619

CHAPTER 4

The Ethics of Explainability

Abstract In this chapter, we turn towards the ethical benefits of explainable technologies. After discussing several instrumental benefits of explainability, we turn towards the deontologically motivated debate of "right to explanation" and their arguments. In refuting that approach on theoretical and practical grounds, we instead introduce the idea of a "claim to expertise", which preserves our practices and allows unexplainable technology to be added to these practices without much disruption. In discussion the consequences of unexplainable technology for democratic governance, this chapter provides some reasons why explainability might be useful politically.

Keywords Explainability · Right to explanation · Expertise · Democracy

The epistemological conditions to know why a technology works a certain way are rather high, and we have little reason to believe currently that we will be able to causally explain certain decision pathways. This has bearings on some ethical questions we might face: intuitively, an unfailing oracle would be acceptable because the performance will outweigh the cognitive purposes of knowing how this performance comes about. Yet, an occasionally failing oracle quickly becomes unacceptable.

© The Author(s), under exclusive license to Springer Nature Switzerland AG 2024
H. Kempt, *(Un)explainable Technology*,
https://doi.org/10.1007/978-3-031-68098-4_4

Philosophers have tackled the question of the ethics of explainability on both an abstract level and in more detail in applied cases. In this chapter, we will first introduce and elaborate on different arguments put forward in favor of strong explainability requirements for unexplainable technology in general before then argue how these arguments all relate to the actual implementation of explainability. As we will see, instrumental reasons for making AI and other unexplainable technologies more explainable do not suffice to motivate strong requirements. Equally, a postulated "right to explanation" may also not provide us with the normative basis to ground explainability in. Instead, this chapter will show that these all fail to take hold on a theoretical level as well as for practical reasons. Lastly, the democratic good of being able to exert collective, legal control over the developers of a technology that will be ubiquitously used in its society can provide sufficient reason for stronger explainability requirements. However, these requirements may be circumvented or fulfilled by other means.

4.1   Why Explainability Matters When It Matters

Rarely anyone in the debate would deny the fact that explainability can contribute some valuable insight into the permissibility of using highly complex machinery. However, at this stage in the development of AI, we can identify two camps that largely disagree on the crucial point of whether explainability is deontologically required for the justified use of some technology, or whether it is not. The answer to this question turns on whether we need to provide certain depths of explanation for technologies involved in decision-making that affect others.

However, there is a practical question attached to a positive answer to the question of deontological necessity: if explainability is deemed to be a normative necessity regardless of its instrumental use, then is this to be understood also as an immediate practical concern, or merely one in principle? Recall our distinction between practical explainability and epistemic explainability: it appears that since many philosophers work on epistemic explainability, the satisfaction of epistemic conditions of explainability (see Chapter 3) is not only the necessary concern but also often the sufficient one, too. The way deontological concerns have been spelled out, however, has stressed the need for practical changes in our treatment of unexplainable technology, especially when implementing them in non-voluntary hierarchical systems. However, before turning towards the deontological demands of explainability, we ought to revise and assess the diverse range of instrumental reasons for why making technology more or fully explainable is desirable.

4.1.1 Instrumental, Utilitarian, and Other Benefits of Explainability

Explainable technology promises to satisfy some of our innate requirement for being in the world. We (at least as the human species at large) wish to understand how things work around us, especially working on us. Whether the explanations are all exhaustive, precise, or with predictive power is not always of high importance. Often enough we simply seek a plausible explanation for something that happened. This applies to events in the world, both natural and cultural, but also includes our own thoughts, other people's actions, and other "basic" occurrences in the world.

When considering technologies, the psychological effect of and desire to "understand" (in the sense of "having an explanation", whatever that explanation may be) becomes visible in tendencies to anthropomorphize such technologies. It is easier to assume that a Roomba-cleaning robot is "hungry" and thus needs to be "fed" with dirt rather than correctly representing the robot as having a technology inside that, somehow through rounds of training and many different sensors is capable of gathering dirt from the ground and is explicitly programmed to do so.

Similar applies to chatbots and their rather impressive skills of presenting human-like behavior to the degree of eerie levels of conversational depth. It simply might be psychologically less costly to "invent" a plausible explanation about the human-like nature of this artificial intelligence than it is to research or—if known—reiterate and represent the structure of large language models and their ultimately well-understood technological skeleton. This borders on the idea of taking an "intentional stance" towards that technology, i.e., that assuming some technology has intentions is less psychologically costly with the same or better explanatory and predictive power for the machine's behavior as the "correct" representation of the going-on in a machine context.

Such a stance could also enable people otherwise fearful of operating complex machines to interact with them, as their own explanations might provide a certain amount of comfort and self-confidence to interact with these machines. Being able to converse with a personal AI assistant in one's "own voice", i.e., without having to switch into the mindset of operating an unexplainable machine, positively affects people's willingness to use the machine's full potential and integrate it into their own lives.

However, explanations not only serve as psychological benefits for orientation or complexity reduction to the degree of regaining one's agency. They also have practical benefits, depending on the kind of explanation. In the following, we are discussing some of these practical elements that are usually put forward in dealing with unexplainable technologies. These are worth considering separately in the debate surrounding explainability as they mark a distinctly different set of reasons for increased insight into explainability.

4.1.1.1   Predictability

A good causal explanation for an event or behavior usually provides a prediction about when such an event or behavior will happen again. This is because a causal chain if without gaps, should provide a determinant answer to questions of what will happen. Successful if–then explanations, thus, are highly desirable, as they enable us to either set the "if" properly to create the desired outcome, or to avoid an undesired outcome by avoiding the if.

Usually even in many natural sciences causal explanations are often made post-hoc rather than predictive, as the complexity of causes makes a determinant prediction about future events virtually impossible. However, this is why statistical models, in which the likelihood of an event occurring, is usually the modus of explaining and predicting events simultaneously. For AI-models, similar appears to apply, as deep neural networks do not allow for causal explanations and thus for a determinant prediction about the exact output it will produce, but merely a statistical approximation about the forms and contents it will take. However, these predictions, despite not being determinant but merely approximating, are still useful.

4.1.1.2   Validation and Technological Debt

The ability to validate and assess an automated decision's reliability is a key demand for the ethical permissibility of automating decisions. Many argue (e.g., Lombrozo, 2011) that explainability is instrumentally valuable for that goal, as it allows a better insight into the workings of the machine and thus allows for easier detection of vulnerabilities, weaknesses, and problem areas.

However, another dimension of this benefit is what has been dubbed "technological debt" in the debate about implementing technology that has not fully matured yet. In short, technological debt describes the increased cost one incurs by using technology that might need fixing or debugging along the way. This cost is multiplied by its use, as the more a technology becomes implemented in important decision-making features, the more it will cost to fix bugs and other issues along the way. Technological debt has been especially problematic with the implementation of large language models because of (a) their fast widespread implementation despite poor reliability and several security and safety concerns, and (b) the challenges of realigning, debugging, and troubleshooting in large models of generative AI. An example here is Google's issues with implementing AI into their search, which has been trained on poor data and thus has been producing highly problematic results. These results need to be hand-pulled from the AI, unless the entire AI-project is to be taken down (Newton, 2024).

Increased explainability here allows for a more precise troubleshooting capacity (see also the point on maintenance below), but generally reduces the tech-debt that digital societies are currently incurring without being fully aware.

4.1.1.3   Maintenance

Technology wears down over time or happens to fail in some unforeseen situations. In complex machines, different parts of the machine will wear down differently and at different paces, requiring different amounts of regular checks and attention. The need to incorporate the dimension of time into the assessment of the functionality of machines gives them either an expiration date or the need for maintenance. However, the better understood the principles behind the machine, the workings of its parts, its material hardware and software qualities, and its wear and tear in different kinds of use, the better the prediction about the need for certain part-replacements or re-adjustments. The practical explainability of a machine is of chief interest here, while pragmatic explainability is equally important—especially considering that systems are often attended to and maintained by people with a certain degree of experience who can detect weaknesses in the machine's constitution based on their many previous encounters with the kind of machine.

Artificial intelligence is no exception in this regard, even though the material wear and tear is less affected by its unexplainable features.

However, the mere software requires maintenance, updates, refinements, and other improvements. Especially "live learning" systems, i.e., those AIs that are being constantly trained while in use to further improve their skills might suffer losses of performance or become more vulnerable to adversarial attacks. Emerging biases and other issues may also emerge only in use that might require a correction or an adjustment. It is a fair assumption that the more explained these systems and applications are, the more likely they will be able to be maintained better. Increasing our understanding of explainable AI is one of the main reasons for computer scientists who are, as stated before, much more motivated by the utility of such knowledge rather than its theoretical, non-practical value.

Approximate or merely "statistical" knowledge about their functions is seemingly sufficient for many purposes (as LLMs are currently in use and are being maintained), but also demonstrates on occasion the limits of a full causal explainable decision-making structure.

4.1.1.4   A Duty to Maintain, a Duty to Explain?

We might argue that the fact that maintenance is a necessary technical action for the continued usability of a technology obligates the provider of said technology to be able to maintain it. The duty to maintain one's technology, then, could constitute a duty to have the technology explainable as otherwise one could not guarantee its proper durability and resistance to issues of wear and tear, emergent biases, and the like. If the provider is not able to repair or maintain the technology's usability, then the provider might violate their duty to stand by their product for a minimal period of time.

Especially in cases where a replacement product is hard to come by, i.e., in highly personalized or specified products (such as AI that is fine-tuned for specific purposes), their breakdown may be accompanied with their irreplaceability. So we might consider that at least irreplaceable technologies ought to be maintainable.

This argument, which would combine the instrumental value of maintenance with the duty to be able to maintain one's technology, could ground explainability requirements. However, even though it would make sense from an economic perspective to keep a technology around without having to replace or reset it every time it breaks down, this is still a plausible, albeit costly, option. Additionally, especially in cases of contemporary AI, some of the fixes for bugs and other issues can be implemented without having the full extent of epistemic explainability at the ready.

Engineers have been able to fix issues, e.g., with an unforeseen "sluggishness" of large language models' responses to user requests, without having a fully detailed explanation for how the language models transform input to output exactly.

The fact that some highly personalized or specified products may be lost once the break is also not a reason to require providers of a technology to be able to explain the technology: first, we should remain skeptical about whether a technology, even an unexplainable one, is ever truly irreplaceable, even with a copy (or whether one some there cannot be a copy), and if so, whether that should still engender a duty to maintain. "Custom-made" products ought to be only as maintainable as the baseline model they are based on.

Ultimately, the sole duty to maintain does not imply a duty to explain, even though being able to maintain properly is desirable and could require high levels of explainability. The limits to which a merely working but not fully understood technology must be maintained might be determined democratically (or, at least, technocratically). That discussion, however, is better suited for the concerns of explainability as a democratic good (see Sect. 4.3).

4.1.1.5   Re-enchantment of the World

Another reason for stronger explainability requirements from an instrumental but potentially non-utilitarian consideration might be that with an ever-growing presence of technology that cannot be explained by anyone (and which is understood to be the case), we might re-enchant the world with entities for which only projected, approximate explanations and best guesses can provide explanations (cf. Musiał's analysis on the matter (Musiał, 2019). As a matter of fact, many people might already be on the path to losing an explanatory toolkit with which they can de-enchant the world. We do not stare into the sky, baffled by the fact that there are giant metal tubes flying through the air (i.e., planes), because we have a rough understanding of how aerodynamics work; similarly, we do not marvel at the same-day delivery of some online store for a product that was assembled just weeks prior on the other side of the planet; all human activity carries an intuitive level of explainability with it. Once we consider robots and AI, of which we are told that even their engineers cannot with certainty predict their next move, their behaviors, or their further developments, it might appear like the world has regained some of its enchantment. While this in itself is not a concern, we might wonder whether some of the long-term effects attract negative consequences. The less explainable an AI becomes, and the more enchanted the world appears as a result, the more space is there for exploitation of those who are overwhelmed or have too little an understanding of modern digital machinery.

We see this in part by AI's presence to cause psychological alignments on the human side (i.e., feeling excluded from a human–machine-conversation if the machine is actively ignoring the human interlocutor (Rosenthal-von der Pütten & Bock, 2023) or anthropomorphism which not only extends human behaviors to machines, but restricts human activities to accommodate machine behavior). These psychological developments might represent a moral risk of treating some machines as if they are more than they are. While this entire debate is pointing towards the debate on moral and social standing, the point here is rather the risk of these misrepresentations being caused by a lack of understanding from the human user side, rather than a paradigm shift in our moral practices and ascriptions of moral status.

This suggests that there are substantial moral reasons to partially work against such a development: the more unexplainable these machines are, the more (phenomenologically) seem these machines to be something appealing on a non-mechanical level. This is also being taken up, in part, in the literature surrounding AI-theology and the philosophical exploration surrounding the potential influence of this technology on religious beliefs and their formation, communication, and performance.

One point to consider here, however, is the question whether this can (solely) be rectified by improved explainability. It appears that enchantment "wears off" once we accept that there is a technology that we might not understand but can replicate on a reliable level: eventually, the rule-guidedness of technology reemerges as a guarantee of human ingenuity at work, not some mystical machine that can things we cannot, does things we do not, and all that without any human being understanding what is going on. It seems unlikely that, once the novelty has worn off, proper legislation has been put in place, the first products have turned out bad, and the hype has worn off, that AI will be a revolutionary, disruptive, exciting, but not re-enchanting technology.

4.1.2 Rights to Explanation?

We now turn from instrumental reasons for assigning value to explainability to deontological accounts of explainability's normativity. The underlying assumption about many conceptualizations of explainability lies with the idea that there is a normatively grounding right or claim that people have in cases of being subject to a decision made by an AI (or to which an AI has contributed in substantial ways). This grounding claim can be characterized in a "right to explanation" which ensures decision subjects that the treatment they receive is explainable to them so their informed self-interest can be preserved. Several accounts have been put forward that aim to establish such a right with different motivations.

From a jurisprudence and philosophy of law perspective, it remains controversial whether or not such a right has been established yet, for example, in the General Data Protection Regulation (GDPR) of the European Union, which came into effect in 2017 (see Wachter et al., 2017 for a negating answer, and Selbst and Powles (2017) for a positive one). Further, there is also a legal debate on whether such a legal right is achieving what it is supposed to accomplish (Edwards & Veale, 2017; Goodman & Flaxman, 2016). While the legal debate is inconclusive, we will turn towards those accounts that propose a right to explanation for ethical reasons. Usually, autonomy, agency, and accountability are the key concepts that motivate these investigations (see, e.g., Vredenburgh (2022), Jongepier and Keymolen (2022), Smart et al. (2020), Rueda et al. (2021) and several more).

4.1.3 A Representative Account of the Right to Explanation

Kate Vredenburgh's account of a right to explanation (2022) represents a strong and detailed proponent. Analyzing and detailing her account should provide a good representation of both the strengths of such accounts and the inherent struggles with the right to explanation, which we will discuss and thus is a useful target for our analysis. Her account is also larger in scope than other, similar accounts as Vredenburgh is arguing from a rules-based domain ("non-voluntary hierarchical systems") rather than practical domains (such as medical decision-making). This is another reason to consider her approach more generalizable over others, and thus a more complete account.

With this larger scope, as it also addresses other unexplainable, non-voluntary, hierarchical systems such as bureaucracy, and unexplainable AI-systems. Thus, it fits within the question of whether unexplainable technology can under specific circumstances represent a violation of one's right to explanation.

She claims that given the fact that we are subject to decision-making procedures that—by the nature of those procedures—remain opaque and unexplainable to us (unless we invest a considerable amount of work to achieve the levels of expertise to appreciate the functions of this system), the decisions imposed upon us by such a system affect our ability to set our own *self-informed goals and self-advocacy,* which come as a cluster of abilities: representation, agency, and accountability (Vredenburgh, 2022, 212–213). As we have a "general and widespread" interest in such self-advocacy against non-voluntary and hierarchical systems that make (or rather: produce) decisions that affect us, and such interest is morally weighty, a constraint ought to be put in place that protects and restores the decision subject's ability to self-advocate. For Vredenburgh, this is realized in the form of a moral right, as a rights-account meets the conditions for such a constraint.

A right to explanation, thus, serves as both an individual's claim against a decision made by an opaque decision-making process, but also as a liberal means to enable citizens to resist decision-making processes that could be misused: in the name of precision or all-encompassing pathways, decisions may grow absurdly complex and thus preclude citizens from exercising their civil rights to control and question those decision-making processes.

Vredenburgh acknowledges that some explanations, if not guaranteed by the systems themselves through the right to explanation, can be performed by expert advocates to a not-too-costly way of reducing complexity without misrepresenting the core of the matter.

Some elements of this account appear similar to a debate that has been advanced in the field of medical AI ethics. The idea of *contestability* (Sect. 2.8), i.e., the ability to contest some of the decisional suggestions made by an AI (see Ploug & Holmes, 2021), requires a large number of explanations for the intended decisions. Especially the origin of a decision, the role of AI in that decision, and the details of that AI can factor into a patient's ability to contest the decisions made are claimed to carry relevant moral weight. However, Ploug and Holmes explicitly limit their account to issues concerning medical treatment.

4.1.4   Other Accounts

A similar account has been put forward by Jongepier and Keymolen (2022), who first reject (as does Vredenburgh) the assumption that automated decisions motivate a right to explanation metaphysically, by virtue of being a fundamentally different kind of decision-making structure than, say, an expert. They base the *epistemic good* of an explanation, however, on the need of decision subjects to be deliberative agents, which requires knowledge-based explanations, and thus reasons for certain decisions. While "deliberative agency" and "ability to self-advocate" both go in a similar direction, Jongepier and Keymolen interpret explanations as an epistemic good for agency, while Vredenburgh presupposes agency as part of self-advocacy and thus sees explanations necessary for determining informed self-advocacy rather than deliberative agency alone.

They specify that a "general right to explanation" must come with some practical ramifications and limits. They call these "epistemic conditions" for deliberative agency. First, they identify that there has to be an institution that is responsible for providing the epistemic goods. Second, the relevant epistemic good must be "substantial", as it should inform our beliefs, which in turn should inform our identity or lifestyle. If both conditions are fulfilled, we ought to consider explanations of the knowledge-producing kind, i.e., a reasons-based explanation, to be a right of a person, as otherwise their deliberative agency could be affected.

Others, like Smart et al. (2020), Munch, Bjerring & Mainz (2024) and Rueda et al. (2021), concentrate on moral justifications and procedural fairness in high-stakes decision-making. Smart et al. argue that a moral justification can only come from human decision-making contributions, and a lack of explainability precludes humans from adding a moral perspective to the decision, even if highly reliable. Munch, Bjerring & Mainz add to this point by focusing on the fact that even low-stakes decisions can engender a right to explanation, making this a strong deontological position.

Rueda et al.'s central claim means that distributive justice is not sufficient for some decision-making circumstances, for example, organ transplant allocations. Even with a highly precise algorithm that can detect all the relevant medical parameters, those who are not given the transplant are owed, procedurally, an explanation that goes beyond the fact that a highly precise machine determined the outcome.

However, just because a "right to explanation" is posited does not mean that there are immediate consequences such as opportunity costs to reject an AI-based decision-making tool. Kawamleh's account (2022) illustrates that we might already cover the right to explanation with currently available tools. In her view, coming from a medical perspective of being made a subject in a medical decision that used AI-powered decision-support systems, informed consensus is the defining and decisive measure to ensure a patient's autonomy. The information necessary for ensuring the ability to informed consent, i.e., preserving the patient's autonomy, must be provided by the machine. However, unlike other accounts (cf. Bjerring & Busch, 2021), Kawamleh argues that contemporary AI-based decision-support systems fulfill these requirements, as they should not be held to higher standards than contemporary practice (see also Zerilli et al. (2019), Kempt et al. (2022)). In her argument, the fact that machines can provide accurate assessments about the reliability of their prognoses (an ability that many human physicians, unfortunately, do not possess) constitutes sufficient information to equip patients with the ability to make informed, reliable choices for themselves.

This does raise the question of whether defenders of a right to explanation actually aim at explainability, or if they merely mean a substantial level of interpretability, i.e., the ability to understand the machine's working without fully being informed about the specifics within the decision-making process. Being offered an interpretation of the behavior of a machine or an explanation of said machine must not necessarily lead to different epistemic capabilities within or outside of clinical decision-making.

Lastly, we ought to mention Colaner's (2022) account in which he provides three non-right-related reasons why explainability is supposed to be inherently valuable. Concentrating on the idea that a lack of explanation would be "dehumanizing", he discerns that lacking explanations for decision subjects are deprived of meaningful control over procedures (mirroring Smart et al.), the second one of not ability to "fight back" (mirroring Ploug and Holms), and a third one of affecting one's dignity, for which he fails to provide any further argument.

4.2   Interim Conclusion

At this stage, we can detect a trend in deontological accounts that centers around a "right to explanation" as a way to protect the autonomy of those being made subject to a decision-making process that is either partially or completely done by an (unexplainable) AI. This appears to also be the tacit premise in many other accounts that argue in favor of stronger explainability requirements: there seems to be a deep-seated intuition, potentially motivated by considerations on epistemic explainability, that we have to have a claim against opaque decision-making structures qua them being opaque (even if they are wholly beneficial, unfailing, etc. See Sect. 1.2 for the intuition-pump of the unfailing oracle). While it is speculative to assume that there are no viable alternatives to normatively ground explainability regardless of its instrumental value, we have yet to see such an alternative emerge. The following argument against higher requirements of explainability of AI of grounds of its opacity, however, should not be affected by other approaches to ground explainability requirements.

4.3   Expertise, Explanation, Trust-Bridges

In this chapter, we are taking the argument from a "right to explanation" seriously. The right to explanation might not stand in the way of using unexplainable technology as it has been put forward before, either by Vredenburgh or others. Exceeding the mere refutation of the ones actually being put forward, like the accounts from Vredenburgh, Jongepier and Keymolen, Smart et al., and others, I argue that the right to explanation is in itself undermining some highly valuable social agreements, i.e., highly specialized expertise. Expertise, however, is a precondition for the wealth and well-being of our modern world. The benefits of this modern world ensures many of our opportunities of self-realization. These benefits rest on the fact that in most areas of human activity, there are experts ensuring the functioning of their respective field (Fricker, 2021).

This, in and of itself, is not in disagreement with a "right to explanation". Most people will agree that highly specialized and siloed expertise is a modern necessity to ensure our way of living. What Vredenburgh and others are pointing at, however, is that in some areas of human activities, the non-voluntary hierarchical systems to which we might be subjected owe us an explanation for their decision-making pathways in order for us to assert our own autonomy.

To clarify why we reject this argument, we need to first elaborate the nature and role of expertise first, and then show how expertise and explainability part at a crucial normative moment. This does not mean that the one replaces the other, but that expertise in general is predicated, as an "expertise ecology" on the lack of explainability demanded from those experts. An "expertise ecology" captures the observation that expertise goes both wide and deep (Fricker, 2021; Goldman, 2001): it goes wide to very many different topics, and comes in many different forms: expertise is not limited to having acquired a certain amount of knowledge in a specific field, but also comes in practical knowledge, experience, skill: artistic expertise, for example, could capture the ability of a pianist to play a certain piece exceptionally well or for a goldsmith to fashion an elaborate wedding ring. However, expertise also goes deep: the specialization of an expert is not limited to their respective general discipline, but often fans out into highly specialized sub-disciplines and sub-sub-disciplines. So much so that experts in neighboring fields may still appreciate the reasoning and can, without great effort, become an expert in that neighboring field, but not without any effort. The depth of expertise produces unique knowledge that can and will affect most people in some form or another. The effort it takes for experts of neighboring disciplines to achieve the level of expertise demonstrates the limits any explanation to non-experts of that area will have.

The "roots of expertise" go deep and wide and form its own ecology in which every part is dependent on each other (with some exception in which expertise might be considered superfluous). It is not only practically impossible for anyone to understand all the other expertise-based decision-making processes (especially in expertise that rests on non-propositional knowledge, like skills and experience), but also undesirable to burden each other with the demand to become an expert in any more than one's own field.

This does not imply that experts cannot explain their judgments to laypeople (who may be experts in their own right in a different area). Most assessments of an expert include some form of ability to break down highly complex processes to an understandable level without losing too many relevant details. However, these explanations do come at a cost: either, the knowledge necessary for laypeople to enable an expert's judgment is very labor-intense, takes time and other resources, and may be virtually impossible to do in a practical way, as it might take years to understand a specific judgment's background. Or the explanation is not gapless in its depth, resulting in the need for trusting the expert in some dimensions of their explanation.

The latter might be the standard case: expert explanations are usually spotty and have gaps of knowledge, not because the expert has knowledge gaps (even though the discipline itself might harbor some gaps as our shared knowledge has not yet advanced to provide some gapless explanations), but because any breakdown of explanations that are gapless are, again, labor-intense and require the layperson to almost become an expert in their own right. If expertise is to work, those gaps have to be bridged with something other than an explanation. This "other" factor, we contend, is trust.

Trust as a mediator between laypeople and gaps in expert opinions, however, is not as expedient as might seem at first: if we still have to trust the expert in some of their recommendations or judgments, while some other parts are explained to us, then the question emerges which parts may be relegated to "trust the expert" and which ones must be explained. Often, the former appears to be the reliance on established science that other experts may also operate from or some highly complicated mathematical principle that they used to arrive at their results. The latter, however, often is required to explain how the judgment in that very instance comes about: What are the concrete matters of the case that render the expert's opinion the opinion that it is?

Determining these informal differences, even when formalized, is a rather difficult undertaking and cannot be expected to be performed by laypersons. Thus, some amount of trust has to be expensed to an expert by virtue of them being an expert on the subject matter. How one achieves the status of expert is not the matter here, as this is part of social institutional negotiations and may vary greatly from society to society and areas of expertise (some expertise might come in the form of formalized educational careers, some exclusively by experience, some other by demonstrated track records, etc.). However, the fact that expert must be trusted for their explanations to work is a key point here: because if trust must be expensed in the face of explanatory gaps, then what is exactly won by bestowing a right to explanation to decision subjects (or rather, a duty to be explainable to the deciders), if some of the explanations will require the decision subject to trust the explainer?

To further complicate things for decision subjects, an expert's judgments, and thus their decisions based on such judgments, are somewhat final to the decision subject (and layperson): not only are laypeople confronted with the contents of a subject matter they have very little knowledge about, they also have very little recourse to test the explanation for correctness. In some highly complex decision-making structures, the decision subject thus not only has a right to explanation (if there is such a thing in the first place) but a right to a proper explanation (see the following chapter for a closer inspection of this argument) with a duty to understand. We might want to consider also the fact that there might be explanations that could satisfy the right, but are returned on the ground of not being easy enough to understand.

In trust-based interactions with experts, such a claim and its associated practical issues do not really factor in. Two cases are illustrative here to show how a right to explanation works different to some expertise, and how the latter works better than a former could, especially in cases as characterized by Vredenburgh.

4.3.1   The Public Defender

We may be afforded a public defender as an established means to combat hierarchical systems of decision-making that laypeople simply would not have the resources to learn and navigate effectively. This is a well-established and justified social solution caused by the insight that a highly complex legal system, which is built on decades of precedent and yet at times changing legal interpretations and rulings, is worth having as a matter of legal certainty, enabling cooperation, and general institutional stability. While we allow anyone to advocate for themselves and themselves alone, the introduction of a public defender acknowledges that self-advocacy in the courtroom is most often a troubled strategy and that some people may not have the means to afford a lawyer that can defend them otherwise.

We thus agree to create hierarchical, non-voluntary systems of highly complex decision-making procedures and accompany that by providing experts who work on our behalf. This is done instead of aiming at making the legal system less complicated and establishing a right to explanation that would enable more often direct self-advocacy.

The system of public defenders works because of our trust in their work increases our ability to self-advocate. This is because the direction in legal defenses is not to achieve a morally just outcome, but to get the advocatee the best possible legal outcome. This also includes that we do not question or doubt the defender's strategy at every step of the way, that we trust their expertise, and that they can trust us in our testimony. The reason we afford each other a public defender is not rooted in a right to explanation; quite the opposite: it is the denial of a right to explanation, as one cannot assert such a right against their defender. It is, however, rooted in the idea that we have to have a tool against the decisions of a hierarchical, non-voluntary system: one that is both achieving our self-advocacy as well as affordable in the sense that it is not unreasonably costly. A claim to expertise, as the institution of a public defender fulfills, is such a tool.

4.3.2 Medical Second Opinions

Medical scenarios are non-voluntary hierarchical systems with patients that are subjects to the decisions made in the system. They are non-voluntary to some degree as the need for medical advice is in many cases essential to human flourishing, and in many emergency only alternative to severe pain and death. We will be discussing the role of unexplainable technology for medical affairs more at-length in Sect. 5.1. However, there is a general discussion about the role of expertise and explanations to be had that is very instructive in our present discussion, comparable to the case of the public defender.

Medical judgments, the decisions based on those judgments, and the interventions following those decisions unfold within a medical context that is following in large parts medical expertise. In many such occasions, the expertise a patient faces resides with one specialized doctor. They provide medical advice, lay out the options, their risks, and thus enable patients to make the best decision for themselves. As we have seen above, informed consensus can be understood as the realization as a right to explanation by charging the "informed" part with a detailed explanation of what is going to happen in a medical procedure and why. However, often enough we grant both physicians as well as patients the opportunity to get a second opinion on a matter. For physicians, this mostly informal process is helpful to solidify their diagnosis and thus minimize risk for both the patient as well as the physician. In fact, some diagnoses require a second opinion automatically (Kempt & Nagel, 2021). For the patient, this process is often formalized, and partly paid for healthcare services:

the patient provides all medical evidence and the conclusion reached by the first opinion to a different physician, to receive a second opinion.

This process is well-established and performed on a regular basis. And yet, it can be rather illuminating about the status of explanations in the field: the idea that an expert needs a second opinion, or that they can be checked by the layperson through getting a second opinion, is remarkable in regard to the status of explanations. Instead of insisting on a right to explanation, which would not support a second opinion (as no explanation may be added, only another expert opinion with lacking explanatory depth), we grant patients (and experts alike) the ability to ask for another expertise. The trust expensed towards experts is not undermined by the lack of explainable decision-making, but rather by the uncertainty of their decision. As we have seen elsewhere (Kempt et al., 2022), most decision-making processes of physicians are not going to provide the explanation needed to satisfy a right. However, instead of this being a problem for medical practice, as the "right to explanation" might have use believe, there are established and justified processes within medical practice that offer expertise instead of a more in-depth explanation. This can be explained by the practical necessities that medical practice often faces: vulnerable decision subjects with potentially limited cognitive abilities to assess their options, a lack of resources to explain medical evidence to patients, the highly complex subject matter of medical sciences, and the siloed medical expertise creating translational barriers all suggest that providing an additional opinion can achieve the same as an explanation.

Next to making a pragmatic case against the right to explanation, this also refutes the idea that an informed consensus must require explainability of the decision-making process: reliable expertise can achieve the normatively required levels of providing decision-making information for decision subjects. An explanation for these decision-making facts need not be included in the explanation.

These two example cases—the public defender and the second opinions of physicians—are suggestive of what we will develop further down as a "claim to expertise", rather than a right to explanation. Before we can do so, we should first discuss the issue of subjective explanatory sensibilities and requirements (4.2.2) as well as the re-introduction of technologies to our argument, since we have thus far barely touched upon the connection between expertise, explanation, and artificial intelligence.

A right to explanation, in opposite to expertise, then, also is accompanied by an at this stage unexplicated duty of understanding. As we will elaborate in the next section, a right to explanation may remain toothless in the light to the required cognitive labor of the decision subjects.

4.4   EXPLANATION SENSITIVITIES

One often overlooked difference between making an AI explainable in the stricter sense and having an expert act as a mediator or advocate is the difference in receptivity and understanding of decision subjects. While the liberal idea of a right appeals to many people interested in being very well-informed about their decisions, it leaves little space for those who choose differently. A right to explanation will inevitably lead to the burden for a proper explanatory appreciation to lie with the decision subject, rather than the decider. This is because not only will there be legal standards emerging for specific AI-applications and their explainability as a means to settle the question of what customers/users should "bring to the table" for AI companies to satisfy these rights, but these standards will also most certainly be used against those with the fewest capabilities to understand their rights.

Rights, as an institutionalized means of providing people with moral and legal claims that can be leveraged in their own interest, are a double edged sword in this regard: they make all customers the same, as it would be highly problematic to posit different depths of explanation in the same right for users/customers, or to posit different explanations depending on the user. In order to not totally wreck the expertise ecology with unreasonable burdens and costs, and recognize the rather obvious point that users do have different requirements for explanations, we fairly quickly are compelled to believe that a right to explanation as a general claim for people subject to technologically unexplainable decisions is too strong a claim.

Enabling informed self-advocacy most likely will still involve some additional actors, i.e., experts, that mediate between the materially redeemed right to explanation, on the one side, and those who are not able to use that level of explanation for self-advocacy. It thus puts into question whether establishing such a right to explanation is the right normative means for the purpose of enabling decision subjects to become or remain able to advocate for their own interests. It also puts into question if this is not merely doubling the social costs required to satisfy such a right.

One argument to defuse these concerns is the differentiation between a general right to explanation and a domain-specific right to explanation. Rather than specifying the explanations we are owed, we could argue that the situations in which such a right is supposed to emerge carry specific features (e.g., when the ability to self-advocate carries a certain importance, or when the decision subject is especially vulnerable (i.e., in medical cases)). In this sense, some situations require a more "engaged" right to explanation, while others may not at all (even though self-advocacy would improve the situation overall). Vredenburgh will probably reject this notion as her definition of a right to explanation is angled against cases in which non-voluntary, hierarchical decision-making procedures are the premise for her argument. However, but even in those situations, paradigmatically "AI systems" and "bureaucratic systems", the stakes of decisions are going to vary considerably, and thus not every explanation has to cover the same moral risk. In this sense, proponents of a "right to explanation" owe us the precision of what kind of explanation we are owed.

While informed self-advocacy sounds like a shared denominator, we ought to be aware that the receptivity and ability to work with a specific explanation from a machine or bureaucratic institution varies greatly among different people, their situations, capacities (see, e.g., Nagel & Reiner, 2013, pointing towards the fact that more information does not necessarily enable more self-advocacy), and, frankly between different AI-systems. Not every AI, for example, can be explained in the same intelligible way, and different AIs will require different amounts of work to come to an explanation we deem to satisfy any such right to explanation. Jongepier and Keymolen, thus, demand "useful explanations" (Jongepier & Keymolen, 2022, 49), which are subject-relative. However, as we have seen above, an individual's right to explanation cannot reasonably be expected to come with an individual explanation (besides the specifics of the case). The principle behind the decision-making pathways remains the same.

We might be able to strengthen an account of the right to explanation by differentiating between the context in which more detailed explanations are necessary (independent of a decision subjects knowledge and ability to work with that knowledge). Still, we may also lose some of its normative appeal on the way: if I only have a right to explanation in cases in which my self-advocacy is deemed to be important enough, then recognizing those cases becomes most salient.

Ultimately, this argumentative move will be rejected by those insisting on the right to explanation being a moral right: the deontological reasoning for such a right is not conditioned on cases in which self-advocacy is important enough; rather, self-advocacy is a condition for permissibly being subjected to the decisions of non-voluntary, hierarchical systems in the first place. However, we have seen that this is not without its pragmatic issues.

4.5   Debunking Explainability through Expertise in Tech: The Case for a Claim to Expertise

Now, where are we in our discussion on the question of why explainability of technology matters when it matters? We have seen that explainability is often useful as an instrument for other, higher goals: to achieve better insights about the reliability of a given technology, to maintain that technology if it wears down over time, to educate human users about the functions of that technology to keep from them falling in anthropomorphic traps or tech-hypes, and so on. We have also seen that there are ambitious arguments about explainability of technology being relevant in and of itself because it is a necessary requirement for preserving autonomy and deliberative agency in technicized and digitized contexts. These requirements come with specifications, as the decisional context must be non-voluntary, sufficiently relevant in consequence or kind, hierarchical (i.e., non-discoursive), etc. This necessary requirement for explainability to preserve autonomy of decision subjects has been dubbed the "right to explanation" that one has against technological decisions (as well as other opaque, non-technical systems).

Our analysis has shown that such a right is causing a number of challenges: it would undermine the expert-ecology in which wide and deep areas and silos of expertise would come under scrutiny to owe decision subjects explanations, where precisely the status of expertise is supposed to replace explanations; it would produce either be a highly costly individualized or a highly generalized, often insufficient explanations, which merely *satisfices* the right for explanation and thus invite gamifications of the right. Almost all explanations still require a fair amount of trust towards experts and expert systems to operate within reasonable, evidence-based bounds, which minimizes the moral gain of having some of the crucial decisional pathways explained to decision subjects.

In this analysis, we have seen that trust thus plays an irreducible role on guaranteeing decision-making processes in which expertise is needed. We have also seen that for some non-tech cases, rather than insisting on an explanation, the established solution has been the grating of even more expertise, which would work as a controlling instance of the expertise involved in the decision in the first place (like in medical opinions), or as working on our behalf against some otherwise opaque decision-maker and their system (like in a public defender situations).

We have, thus far, deduced that there is a claim to expertise, i.e., a morally justified claim to expertise in decision-making structures in which one is subject to a decision. At this stage, we aim to develop this claim to expertise further, by testing it against environments and decision-making structures that fully or partially involve an AI (or any other unexplainable technology). As we have stated expertise is a trust-based, socially accepted division of labor to provide each other with an otherwise impossible level of well-being, the consequences of such expertise is not to make one's decisions explainable.

A claim to expertise for such moments would require the AI engineers (or rather companies) to create partially interpretable, auditable, and other models. These, however, are not meant to be explaining the decision to the decision subject but rather make the decision assessable by experts—those very experts that will protect our decision-making capabilities by not providing explanations but expert opinions.

Depending on the decision situation (as we differentiate in Chapter 5), the claim of a decision subject to expertise can feature in different ways. Similar to a public defender, we might consider decisions that affect our civil rights, finances, and other administrative-political standing (loan and job applications, parole, etc.) to be in need of an expert that can assess the decision and the reasoning behind that. Equally, decisions that affect us directly on a personal level, like medical decisions, may only be administered if we were provided the expertise of human doctors who have access to the automated decision (see Kempt & Nagel, 2022). Having a doctor at one's side who can interpret the machine's behavior, i.e., its decision, can provide a more trusted and morally grounded assessment of the trustworthiness of a machine. This is to be seen in contrast to the provisions that a machine may provide explanations without the relation of a medical expert. The explanation of a machine simply stands on its own, and might even be claimed to be sufficient to allow machines to automate entire diagnostic and treatment processes if they are presumably "explanatory" enough.

We argue that an expert who has a working interpretation of the machinery but only approximate knowledge (as there is no expert that possesses sufficient knowledge), who is committed to making decisions in the layperson's best interest not only provides the moral demands on the same level as an explanation would, but is (a) closer to our contemporary actual practices of how we deal with a virtually (or actually) unexplainable decision, (b) less costly, as the technology would not have to be changed, (c) utilizes the currently available technology without requiring ever-growing models to adhere to explainability standards, and (d) captures many responsibility questions via second-opinion-style assurances.

As the claim to expertise has been realized in different instances before, we have reason to assume that integrating this claim into policy of technologically supported decision-making will be minimally disruptive, while taking advantage of the current (and anticipated) technologies without having to wait for them to be made more explainable or weigh their unexplainability against their current benefits. The key benefit of a claim to expertise, if fulfilled, however, lies in the fact that experts will bear some of the responsibility for decisions made in their client's behalf. An explanation, in turn, would either load all responsibility on the decision subject, as they have to deal with the (now explained) decision or, if the decision (or the explanation of it) is faulty or misrepresenting, opens up responsibility questions.

However, we may notice that the right to explanation, and a realization through explainable AI, has two key benefits over expertise claims: expertise may be abused in what has been called "expert-manipulation" (cf. Guerrero, 2017), in which an expert exploits their epistemic domain-specific superiority to manipulate laypeople to the expert's favor. Presumably, an AI's explanations would not be used to manipulate. However, we can counter this concern by pointing towards the need for incentives for expert opinions that undermine the likelihood of such manipulation occurring, by keeping the access to experts as low as possible. The claim is, after all, to expertise, not to a single expert. We should grant anyone being subjected to non-voluntary, substantial, hierarchical decisions access to expertise that can interpret that decision.

However, this way the second issue becomes even more salient: once the decision-making pathways of a machine are sufficiently granularly explained, they stay that way. Explainable technology is a principle, lasting answer to the problem of informed self-advocacy, while expertise is depending on structural and personal availability. There might, simply put, not be enough experts around to ensure that anyone who is subjected to a decision made in parts by an AI can have an expert advocate work on their behalf.

This is a concern that can be answered in two ways. On the one side, it is doubtful whether there are structural limits to having experts advocate on someone's behalf, and if there are such structural limits, it might just provide a normative reason to limit some automation: if the automation of, say, a loan application is fully automated to the degree that there are not enough experts to challenge the decisions, then the automation might have to be limited or supplemented with the proper expertise-infrastructure to be permissible.

On the other side, this benefit illustrates a general shortcoming in the debate about standards of AI, and thus even strengthens some higher-level justice considerations: the fact that expertise is fragile, contextual, and often undersupplied (especially in cases of medical expertise (cf. Freyer & Kempt, 2023)) affirms the need for more justly distributed expertise worldwide. Take the example of health-insecure collectives, i.e., communities that are lacking medical supplies and support. If we could deliver wholly unexplainable, but reliable medical technology to those places, neither the claim to expertise nor the right to explanation would be satisfied (see, e.g., Penu et al. (2021) how the GDPR affects businesses in Sub-Saharan Africa in their adoption of unexplainable AI).

However, a lack of expertise to fulfill a claim of a community supplied with unexplainable technologies will only increase the need for them to also be supplied with expertise: the technology itself remains untouched as a moral concern, but the circumstances of its use. If a right to an explanation were to be implemented that can work without expertise (as we have seen, if it cannot work without trusted expertise, then there is little gained over a claim to expertise), then we might wonder if, and if yes, under which limited circumstances unexplainable technologies may be used to be distributed and implemented.

4.6   Further Considerations

Trust, as a key to inter-human recognition, recommendation, and implementation of each other's expertise (Kästner et al., 2021), is a condition for some cooperative actions. Without these kinds of trusted relationships, not only would the cooperative impetus be missing for people to work with each other efficiently: the constant control of each other essentially breaks down any efficient way to share the labor necessary for a larger, shared benefit; it would also diminish the ability to recognize, plan, and effectively implement cooperative actions.

Trust between people, and their recognition of each other's expertise, thus also enables the use of otherwise unexplainable technologies: those who work with the machine, and their ability to interpret the outcome based on their expertise, enable laypeople to make fully informed decisions based on the expert's recommendation or confirmation of the algorithmic decision.

The solution to a right to explanation, either general or specific, could then be the claim to expertise or expert advocates on one's behalf. Of course, this has some drawbacks as well, but captures the arguments presented here quite well: it is not about explanations of a dubious level of specificity that may or may not satisfy a decision subject required explanations, but the strengthening of social requirements for being heard and defended by an expert. This fully embeds unexplainable technology in a social setting of expertise, with all established norms in place.

One practical question emerges here, however: How do we guarantee that everyone has access to expertise, even in situations in which someone may encounter unknown unknowns: we might be treated by an AI and have no idea that we might have a right to expertise that can help us figure out whether the treatment was justified or fair: how does the notion that a decision-making procedure from an AI (or a bureaucratic process or the like) should notify an expert differ from the fact that such a procedure owes us an explanation on its own? It remains an open question, largely targeting the transparency and labeling of decisions. This can be done by making it obligatory to mark any decision that had unexplainable technology take over a significant part of the decision-making process.

Such transparency demands, however, require a different discourse on the governing of AI, which we will turn to next.

4.7   GOVERNING THE UNEXPLAINABLE

Explainability, as elaborated upon in 2.1, is a principle question about certain kinds of technology. It is thus not "merely" about someone's actual access to explanations that make the workings of a machine clear;

rather, it is about whether we are epistemologically, if not ontologically, capable of explaining why some technology behaved the way it behaved. Such a definition then takes explainability considerations out of the public debate, or rather delegates it to more abstract legal or technocratic debates. However, in practice some of the technologies in question are not only unexplainable to most of its user-base, but even to most of those involved in dealing with that user-base. This is not always based on a principle lack of ability, but due to a lack of knowledge within the user-base and their networks. We ought to consider communities that rely on technology with no way of knowing how this technology works somewhat vulnerable to the potential adverse effects of this technology. Democratic control, then, becomes an ever more important element in reflecting on the ethical permissibility of some technological innovations that are virtually or by principle unexplainable (see also Danks (2022) for a similar account). In the following, we discuss why these digital innovations are problematic from the perspective of democratic control.

First, many of the most relevant technological innovations and ubiquitous adoptions are in the hands of incredibly large, multi-national corporations. And while this is true for other industries, such as airplane manufacturers, diamond miners, or freight ships, the dominating corporations (mostly from Silicon Valley) have a large consumer-facing business model. Websites and apps that utilize recommender apps to suggest us consumer goods, videos, music, news, other people's opinions, and other content form our view of the world; gadget producers such as smartphones, home pods, headphones, and other smart tech influence our lifestyle and expectations about services, functionality, and design of technology. They present themselves to us via marketing, they have power over political agenda-setting and are an essential medium between us and the world (at least in the digital sphere).

All these corporations have a substantial yet subtle power over our lives in different ways. Corporate secrets and technology, which is somewhat the legally sanctioned "unexplainable" technology, are usually key to keep competitors and regulators at bay (see for analyses Burrell, 2016; Rudin, 2019). Unexplained technology, in this sense, might ensure their enduring dominating position in the marketplace. In turn, explainability, understood as a democratic exercise in technology-control, could provide the underlying reason and normative justification for controlling these corporations to a degree they might not like. While this will affect all corporations engaging in AI-development, this is not a means to reorganize an otherwise solidified market. It is, however, a potential re-assertment of the primacy of the state over the technology-companies, which has some beneficial ramifications for the future of AI-development: finding and setting a standard for explainability (see, e.g., Floridi & Cowls, 2019) could re-introduce some competitive playing field, as this would be a challenge for every AI company involved.

To what degree explainability concerns should factor in should be left up to the legislating bodies themselves, as there might be varying differences in what degree the decision pathways need to be regulated in a trade-off with the services these corporations offer. Especially in the medical field, this weighing process represents a delicate affair between the interests of corporations, health-insecure communities, and health-secure communities (see also Sect. 5.1).

However, leaving unexplainable technology in the hands of a few large, profit-oriented corporations may represent a risk for democratic institutions due to an unjustifiable power imbalance between the corporations and the democratic societies these corporations operate in. Explainability demands may decrease these imbalances.

Second, next to the market-position argument, we should also contend that it can weaken our institutions' effective protection of its citizens (see, e.g., what Berry (2021) called "explanatory politics"). The task of democratic control, ultimately, is not exhausted by ensuring the proper functioning of the market but also by providing effective regulations that set safety and security standards. Explainability might be considered helpful in this case, since making opaque technology more explainable will improve the understanding of lawmakers (and thus, in theory, their lawmaking) and of their citizens.

If lawmakers are providing legal pathways for AI to become so embedded in high-stakes decision-making processes that citizens of that legal realm are not voluntarily subjecting themselves to these decisions, it is upon the lawmakers (or rather, the political administration at large) to ensure that the decisions are relatable to the decision subjects. We have seen that this does not establish a right to explanation, but it comes with a claim of citizens against their lawmakers to ensure proper protection from biased or otherwise unfair decision-making. This, lawmakers seem to agree, necessitates some degree of explainability. Thus, we may want to think not in terms of the "right to explanation" of a citizen, but as a "duty to explain" of those making decisions. This can, as pointed out in the previous chapter, be fulfilled by a mediating expert, however, which could ensure the use of cutting-edge technologies.

The concern here lies with the lack of knowledge and resources to provide citizens with such a critical voice, as the decision made might not even be understood to be contestable from an expert opinion. In this sense, laypeople may encounter unknown unknowns in their interactions with highly complex and opaque decision-making systems.

Third, as discussed in a previous chapter on an argument for a duty to some explainability due to a duty to be able to maintain the AI, this could be realized as a legal duty based on some general legal and moral considerations about user protection (in the case of a customer-facing AI) or in cases of high stakes (in cases medical AI, for example). It appears unacceptable from a societal perspective to have an ever-growing amount of technology present in our midst that remains essentially unexplained to the authorities responsible for ensuring our safety (see, e.g., Danaher's point of this constituting a threat of algocracy, where he especially points towards opacity as an issue (Danaher, 2016, 249)). While we have seen this with other areas of technological innovation, e.g., pharmaceutical products, highly complex aircraft software (think of Boeing's unfortunate MCAS system) appears to be in dire need of explainability to be able to be licensed or certified. In this sense, while we have seen that in other cases certifiability can be limited to accuracy and reliability considerations that are determinable by validation of the AI's work, the fact that some issues may have to be fixed in-use, i.e., with quick fixes, or that they may emerge during continued use, i.e., as emergent biases, can give lawmakers sufficient reasons to insist on deeper explainability levels.

Fourthly and lastly, we might want to consider a more fundamental reason one could have that is related to the previously discussed "right to explanation". While such proposals are usually discussed as a right for those more or less immediately affected by the decisions of an AI, that is from a perspective presupposing personal affectedness, the argument here could come from a democracy-theoretic, impersonal one (see Selbst & Powels, 2017): one could argue that citizens of a democratic state have the right to explanation not to guarantee their own autonomy in cases of being subjected to the AI's decision, but merely to ensure that the technology used will be explainable (see also de Fine Licht and de Fine Licht (2020) arguing for explainability as a way to ensure legitimacy of algorithmic decisions). A civil right, rather than a moral one, would take up all other reasons discussed above and would give lawmakers a well-defined tool to discern whether a company is complying with any legislation in this field and act accordingly. In this sense, opaque decision-making structures are generally intolerable in an otherwise open society, even if there is a certainty that these structures remain in democratic hands. While some of these decision-making structures are covert (i.e., on national defense matters), they remain in democratically legitimized institutions. It is unlikely that the engineering of AI will ever be able to claim legitimacy to justify the levels of unexplainability expected from a technology that is nearing infrastructure levels of importance and ubiquity in an open society.

4.7.1   AI Act

Recent legislation by the EU, known as the AI Act (AIA, 2024), has put forward a slew of regulations on how AI may be used, under what conditions it can be developed, and implemented, and what needs to be provided to exert democratic control. Experts argue whether other, earlier laws have established such a right at least in kind or implicitly (see Selbst and Powles (2017) who argue that the GDPR's articles 13–15 establish such a right (GDPR, 2017)), and what size and to what explicitness the AI Act should contain references to a "right to explanation" *expressis verbis*. In the earlier stages of the process, explainability was on the one side feared to play an outsized role in legislative regulation, and on the other hope to achieve compliance with other associated safety and reliability rules within the act. In its final version, the AIA does not include any substantial, direct requirements for explainability, as the principle of transparency (see Sect. 2.7) and the demand for human oversight have been strengthened instead. While there are advocates for furthering explainability efforts within legislation to harness their instrumental values of better risk-assessment and safety architectures (Pavlidis, 2024), it appears that explainability has not been asserted as a deontologically grounded principle. Democratic decision-makers factor in many different reasons to consider, ideally converging on a well-balanced mix that represents the best interest of the majority of people. That means that the omission of explainability requirements in the AIA is not a philosophical position, but will inform some of the intuitions about the need of some moral principles for their political justification and liability questions. In this regard, the lack of explainability in the AIA might have a lasting influence on the issue we might take with unexplainable technology.

4.8   EXPLAINABILITY AND ETHICS: ALL ABOUT STANDARDS?

Ultimately, we can interpret the struggle for the levels of explanations required for the individual and collective acceptability of highly opaque technological systems to be representative of the acceptability of current AI technology at large. This motivates why explainability and all its associated forms, such as interpretability, auditability, and contestability, are of such high relevance for current ethical debates. However, as a powerful technology that will affect our lives in many different dimensions, the setting of standards is the challenge moral and legal experts, and society at large, face now. We can summarize the debate presented here in this chapter as a struggle about the determination of standards for a technology that will change society.

Standards for the depth of explanations of AI need to be settled between several different factors and in different stages, which makes the determination not an easy task. First, we ought to decide if our current standards and explanation practices are of any relevance to the expectations we have for mechanistic explanations. Second, we ought to decide whether any such standards are to be understood "merely" as moral standards or whether they should have an explicit influence on lawmaking. Third, we must weigh the good of explainability against the other goods of a technology that promises to revolutionize, or at least considerably improve, many processes that are currently limited by human, social, or financial capacities. Without depressing research and innovation in this field, societies need to come to an informed position how they want to further develop the role of technology in their communities—especially those technologies that can make automated decisions and that are, currently, unexplainable.

REFERENCES

AI Act (2024). https://artificialintelligenceact.eu/de/ (last accessed May 31st 2024).

Berry, D. M. (2021). Explanatory publics. Explainability and democratic thought. In B. Balaskas & C. Rito (Eds.), *Fabricating publics: The dissemination of culture in the post-truth era*. Open Humanities Press.

Bjerring, J. C., & Busch, J. (2021). Artificial intelligence and patient-centered decision-making. *Philosophy & Technology, 34*(2), 349–371. https://doi.org/10.1007/s13347-019-00391-6

Burrell, J. (2016). How the Machine 'thinks': Understanding opacity in machine learning algorithms. *Big Data & Society, 3*(1), 1–12.

Colaner, N. (2022). Is explainable artificial intelligence intrinsically valuable? *AI & SOCIETY, 37*, 231–238. https://doi.org/10.1007/s00146-021-01184-2

Danks, D. (2022). Governance via explainability. In Justin B. Bullock, Yu-Che Chen, Johannes Himmelreich, Valerie M. Hudson, Anton Korinek, Matthew M. Young & Baobao Zhang (Eds.), *The oxford handbook of AI governance*. Oxford University Press.

de Fine Licht, K., & de Fine Licht, J. (2020). Artificial intelligence, transparency, and public decision-making: Why explanations are key when trying to produce perceived legitimacy. *AI & Society, 1–10*. https://doi.org/10.1007/s00146-020-00960-w

Danaher, J. (2016). The threat of algocracy: Reality, resistance and accommodation. *Philos. Technol., 29*, 245–268. https://doi.org/10.1007/s13347-015-0211-1

Edwards, L., & Veale, M. (2017). Slave to the algorithm? Why a "right to explanation" Is probably not the remedy you are looking for. *Duke Law & Technology Review, 16*, 18–84. https://doi.org/10.2139/ssrn.2972855

Floridi, L., & Cowls, J. (2019). A unified framework of five principles for AI in society. *Harvard Data Science Review, 1*(1). https://doi.org/10.1162/99608f92.8cd550d1

Freyer, N., & Kempt, H. (2023). AI-DSS in healthcare and their power over health-insecure collectives. In H. Bhakuni & L. Miotto (Eds.), *Justice in Global Health* (pp. 38–55). Routledge.

Fricker, E. (2021). Epistemic self-governance and trusting the word of others: Is there a conflict? In J. Matheson & K. Lougheed (Eds.), *Epistemic autonomy* (pp. 323–342). Routledge.

General Data Protection Regulation of the EU (GDPR) (2016). 679. https://gdpr-info.eu/

Goldman, A. I. (2001). Experts: Which ones should you trust? In *Philosophy and Phenomenological Research, 63*(1), 85–110.

Goodman, B., & Flaxman, S. (2016). *European union regulations on algorithmic decision-making and a 'right to explanation'*. https://arxiv.org/abs/1606.08813

Guerrero, A. A. (2017). Living with ignorance in a world of experts. In Rik Peels (Ed.), *Perspectives on ignorance from moral and social philosophy* (pp. 156–185). Routledge.

Jongepier, F., & Keymolen, E. (2022). Explanation and agency: Exploring the normative- epistemic landscape of the "Right to explanation". *Ethics Inf Technol 24*, 49. https://doi.org/10.1007/s10676-022-09654-x

Kästner, L., Langer, M., Lazar, V., Schomäcker, A., Speith, T., & Sterz, S. (2021). On the relation of trust and explainability: Why to engineer for trustworthiness. In *IEEE 29th International Requirements Engineering Conference Workshops (REW)* (pp. 169–175). IEEE. https://doi.org/10.1109/REW 53955.2021.00031

Kawamleh, S. (2022). Against explainability requirements for ethical artificial intelligence in health care. *AI and Ethics., 29*, 1–6.

Kempt, H., & Nagel, S. K. (2021). Responsibility, second opinions and peer-disagreement: Ethical and epistemological challenges of using AI in clinical diagnostic contexts. *Journal of Medical Ethics, 48*, 222–229. https://doi.org/10.1136/medethics-2021-107440

Kempt, H., Heilinger, J. C., & Nagel, S. K. (2022). Relative explainability and double standards in medical decision-making. *Ethics and Information Technology, 24*(20). https://doi.org/10.1007/s10676-022-09646-x

Lombrozo, T. (2011). The instrumental value of explanations. *Philosophy Compass, 6*(8), 539–551. https://doi.org/10.1111/j.1747-9991.2011.004 13.x

Munch, L. A., Bjerring, J. C., & Mainz, J. T. (2024). Algorithmic decision-making: The right to explanation and the significance of stakes. *Big Data & Society, 11*(1). https://doi.org/10.1177/20539517231222872

Musiał, M. (2019). *Enchanting robots. Intimacy, magic, and technology*. Springer.

Nagel, S. K., & Reiner, P. B. (2013). Autonomy support to foster individuals' flourishing. *The American Journal of Bioethics, 6*, 36–37.

Newton, C. (2024). *Google's AI search setback*. https://www.platformer.news/google-ai-overviews-eat-rocks-glue-pizza/ (last accessed May 31st 2024).

Pavlidis, G. (2024). Unlocking the black box: Analysing the EU artificial intelligence act's framework for explainability in AI. *Law, Innovation and Technology, 16*(1), 293–308. https://doi.org/10.1080/17579961.2024.231 3795

Penu, O. K. A., Boateng, R., & Owusu, A. (2021). Towards explainable AI(xAI): Determining the factors for firms' Adoption and use of xAI in Sub-Saharan Africa (2021). *AMCIS 2021 TREOs. 35*. https://aisel.aisnet.org/treos_amc is2021/35

Ploug, T., & Holm, S. (2020). The four dimensions of contestable AI diagnostics—A patient-centric approach to explainable AI. *Artificial Intelligence in Medicine, 107*, 101901.

Rosenthal-von der Pütten, A., & Bock, N. (2023). Seriously, what did one robot say to the other? Being left out from communication by robots causes feelings of social exclusion. *Human-Machine Communication, 6,* 117–134. https://doi.org/10.30658/hmc.6.7

Rudin, C. (2019). Stop explaining black box machine learning models for high stakes decisions and use interpretable models instead. *Nature Machine Intelligence., 1*(5), 206–215.

Rueda, J., Rodríguez, J. D., Jounou, I. P., Hortal-Carmona, J., Ausín, T., & Rodríguez-Arias, D. (2022). 'Just' accuracy? Procedural fairness demands explainability in AI-based medical resource allocations. In *AI & Society, 39*(3), 1–12.

Selbst, A., & Powles, J. (2017). Meaningful information and the right to explanation. *International Data Privacy Law, 7*(4), 233–242. https://doi.org/10.1093/idpl/ipx022

Smart, A., James, L., Hutchinson, B., Wu, S., & Vallor, S. (2020). Why reliabilism is not enough: Epistemic and moral justification in machine learning. In *Proceedings of the AAAI/ACM conference on AI, ethics, and society* (pp. 372–377). https://doi.org/ https://doi.org/10.1145/3375627.3375866

Vredenburgh, K. (2022). "The right to explanation. *In the Journal of Political Philosophy, 30*(2), 209–229.

Wachter, S., Mittelstadt, B., & Floridi, L. (2017). Why a right to explanation of automated decision-making does not exist in the general data protection regulation. *International Data Privacy Law, 7*(2), 76–99. https://doi.org/10.1093/idpl/ipx005

Zerilli, J., Knott, A., Maclaurin, J., & Gavaghan, C. (2019). Transparency in algorithmic and human decision-making: Is there a double standard? *Philosophy & Technology, 32*(4), 661–683. https://doi.org/10.1007/s13347-018-0330-6

CHAPTER 5

Applied Cases

Abstract Assessing four different applied cases where unexplainable technologies are currently in use—medical diagnostics, self-driving cars and AWS, Recommender algorithms, and generative AI. All these applied cases have different ethical conundrums associated with their unexplainability, but none of them require them to be more explainable: whether increased transparency, auditability, or certifiability of their risks, there are ways of responsibly using this technology as it currently is.

Keywords Explainability · Medical diagnostics · Self-driving cars · Recommender algorithms · Generative AI

In the following, we are turning our attention to some of the most pertinent areas in which the levels of explanation of AI are discussed as a concern for their permissible use. These areas still are merely a selection, but represent the variety in which explainability, the stakes associated with the decision-making pathways, and the pre- and post-hoc evaluation of these decisions are relevant. While medicine and automated weapons and cars carry high relevance both before and after the decision was made, especially for individuals involved, recommender algorithms and generative AI-applications may gain moral relevance in their explainability through their ubiquity and cumulative issue manipulative influence on users.

Parsing the difference in these cases is thus a necessity for the proper assessment of the necessary levels of explanation involved. The difference in cases also illustrates that unexplainable technologies in question, even when all unexplainable due to the same construction paradigm (i.e., machine learning), all have different requirements for their explainability if any at all.

By applying the claim to expertise to these four applied areas in which unexplainable technology may be used, we demonstrate how trusted expertise is the key to the justified use of these technologies, and that their implementation ought to not be held up due to their unexplainable decision pathways.

5.1   MEDICINE

One area of application that has received a large amount of attention is the use of AI for medical purposes. AI promises to—if not revolutionize—considerably change medical practice by making the diagnostic and therapeutic-interventional processes more automated. Several reasons have been put forward arguing that implementing AI in medical practice ought to be limited based on the concerns of unexplainable elements of the technology for our healthcare practices and standards (see, e.g., Amann et al., 2020; Kundu, 2021). As a matter of fact, many accounts surrounding the right to explanation or other strong explainability requirements are motivated by medical AI (see, e.g., Bjerring & Busch, 2020, Braun et al., 2020, Grote & Berens, 2020, Rueda et al., 2021, Rudin, 2019, Theunissen & Browning, 2022).

First, the fact that this is a high-stakes environment requires higher standards for both the technology as well as the experts involved. Their judgments can cause or prevent small or great suffering, or even death. The subjects of these decisions are usually in a mentally and physically vulnerable situation, as their pain may influence their decision-making capacities on several levels. Thus, clear and intelligible access to all necessary information to make a decision for an intervention that might cause them great harm or will cause them great harm if they refuse it is a necessary condition for their use. This argument transfers to other areas of automated decision-making, if the stakes are sufficiently high, but becomes especially concerning with medical contexts as the expansion of our medical abilities is of the highest shared interest.

Second, the difference of cognitive positions between medical professionals and patients suggests that the former bear some responsibility towards the latter that may be violated if the decision-making process includes systems that are unexplainable even to the medical professionals. A "duty to explain" might simply be the other side of the coin of a "right to explanation". Failing this duty might harm the role of medical experts as both ultimate bearers of responsibility but also as mediators of medical knowledge towards patients. We might want to consider here that the explainability of the tools in use is a necessary condition for their justified use as otherwise the responsibility practices in medical care break down (see Wadden, 2022). A medical doctor, as goes the argument, simply cannot perform their job morally if they are not possessing a certain level of control and knowledge. This level is unattainable in using medical AI, even if the healing process is successful and reliably so.

Third, patient autonomy. It is supported largely by the ability to informedly consent to certain procedures based on explained findings, the risks and chances involved in the options, and the reasons for physicians to recommend some procedures over others. Introducing unexplainable technology to this could be a cause of concern not only for responsibility questions (as pointed out above) but about the ability of a medical professional to outline the reasons and risks for a certain procedure or the underlying diagnosis. In a "patient-centered approach", these arguments are focused on the interest of the patients, who may be best treated by human doctors who use machines that leave no explanatory gaps (see, e.g., Bjerring and Busch (2021), Grote and Berens (2020), Price (2015)).

Fourth, a consequentialist concern for unexplainable technologies lies with deskilling and overreliance. As we have seen elsewhere, instrumental reasons for explainability, i.e., those in which an explainable technology serves higher purposes, are rather common. In medical practice, the concern becomes salient as some skills that need to be refreshed and used regularly to maintain those skills may become replaced by an automated process. Currently, there is no regulatory consensus on which skills ought to be kept and which may be permanently delegated to automation. This creates uncertainty and keeps the advancement of the education of doctors unguided and potentially patchy, leading to the loss of some important skills while retaining some that may be permanently obsolete.

A more explainable AI may be able to resolve some of these concerns by keeping the knowledge of certain decisional processes around.

These points appear to support a strong explainability requirement for medical AI, on several different levels of application. However, as the term "relative explainability" points to a rule of equality of explanations between humans and machines, we may be able to assess whether the machines in question are actually failing at these points. As we have pointed out, the clinical decision-making process has an ideal, epistemologically justified version and a non-ideal, still normatively justified version in practice. The practice of decision-making in clinical contexts has to incorporate limited knowledge, limited skill, limited resources of time, material, and instruments, limited patient cooperation and mental capacity. Generally speaking, clinical professionals are justified in their use of heuristics and other decisional shortcuts that they might not be able to epistemically and practically explain. Some doctor's gut feeling, based on their experience which, in turn, is based on hundreds of cases of which the doctor may not recall the details, can be a highly precise way of coming to a reliable decision.

Bounded rationality, in this vein, "gets us there". There are two arguments to be had that a transfer of these pragmatic standards to machines is justifiable under certain conditions.

First, we take doctors' standards to be acceptable in their current stage. Thus, as long as we do not go below the established standards, one could argue that this is justified. It is relevant, however, whether these standards are brought on by an otherwise understaffed medical industry (necessitating decisions that cannot be completed by the ideal, rules-based procedures that are epistemically justified) or due to the limits of current medical knowledge and skill: if we can improve the practical contexts of medical decision-making, i.e., increase the staffing, provide more high-quality experts, etc., we should also expect the reasoning process to become more rules-based, including when introducing AI to the decision-making process. However, some decisions may always be based on a doctor's assessment that they cannot quite pin down to the level of explainability some appear to demand of an AI. In these cases, where not the practical context, but the epistemic context limits the explainability of human decision-making processes (while they are still justified), introducing higher standards for medical AI appears to introduce a double standard. Such a double standard, however, is harmful to the success of implementing AI ubiquitously. The moral imperative to continuously improve our healthcare capacities may be hampered by introducing these higher explainability standards for a technology that is outperforming doctors on other benchmarks. Additionally, it may even help to increase the overall performance of doctors if they were allowed to fully incorporate these machines in their clinical decision-making.

Second, the variation in performance and explanations in human doctors simply precludes setting a clear standard for an AI-based algorithm that, presumably, will be used in many different clinical contexts. The question even of an "equal standard" for human doctors implies that there is a baseline of explainability that simply does not exist in medical practice around the world. While there are some practical standards, even those may be undercut in areas of scarcity of expertise or instruments. It appears not only unreasonable but rather problematic from a perspective of justice to establish strong explainability standards for medical AI if such a standard was to limit the implementation of potentially life-saving medical equipment.

The concern of introducing double standards and their harm for patients stands in stark contrast to the "patient-centered approach" in which explainability is strengthened as a principle of biomedical ethics. Some even discuss explainability as a "fifth principle" (Ursin et al., 2021) of said ethics, next to benevolence, non-maleficence, fairness/justice, and autonomy (Beauchamp & Childress, 2001). What is missing in these patient-centered debates, however, is the practical assessment not from the ideal (or best available) medical care, but from the standards of actual patient experience and demands. If, in order to be treated in the first place, we must allow doctors to base some of their decisions on their previous experience and gut feelings, then we ought to allow them to do so. If we want to be treated with the most reliable human–machine cooperative unit in medical practice, we may extend our trust towards the medical professionals operating within this cooperative unit even if they cannot explain how the machine element of that unit operates.

We propose to set standards of explainability, both for industrial work and for patient expectations, along the lines of "relative explainability" (see Sect. 2.2). We elaborate on the comparison in Kempt et al. (2022) based on the considerations about the practical circumstances of medical decision-making in clinics, and discuss the potentially negative role of explainability for medical practice worldwide in Kempt et al. (2022). Relative explainability, i.e., relative to a set of standards of common human explainability of decisions, allows for a differentiated approach to medical decisions.

It is not explainability, then, but a complementing account of interpretability for doctors (Sect. 2.3.) and certifiability for the reliability of the machine (Sect. 2.9) that can provide the necessary standards. Especially delegating the standards of machines to the lawmaking institutions through certification processes guarantees the appropriate levels of reliability not being thought up by philosophers coming from ideal circumstances but by the given needs of a medical community. It also allows for medical professionals to use their common sense and experience, their gut feelings, as well as their specified medical knowledge to assess a machine's suggestion, through interpretable methods. Interpretability requirements are, as we have discussed, lower in their epistemic rigidity, as they pursue different purposes from those of explainability.

The Claim to Expertise in Medical AI: Changing Expectations and Expertise

Patients are not subjected to the decisions of the machines on their own (or at least, we find other reasons than the machine's unexplainability to be decisive here), and thus we should expect that a medical professional with a rudimentary understanding of the technology in question will remain the mediator and delegator of medical knowledge. If that is the case, however, then we should be more focused on making/keeping medical AI interpretable for medical professionals, in the sense that they can glean some of the mid-level processes of the decision-making pathways, can ensure common-sensical results, and thus act as a mediator of medical decision-making and holder of responsibility.

As we have seen, we can compensate for the increasing lack of explainability in high-stakes contexts like medical decision-making by providing interpretable machines that allow a medical expert to read out the decisions based on the evidence available, and certifiability of the performance of the machine overall (given the appropriate political conditions to do such). This does cover the practical and most cognitive purposes we usually have when making expert decisions.

One important consequence of demanding a claim to expertise in medical decision-making is the changing role of medical professionals, which ought to be guided and reflected upon. Sand et al. (2021) pointed towards the need to determine to what degree interpretability (or even explainability) ought to be part of medical experts' EPA (entrustable professional activities). This could feasibly be done via certification and training of the medical personnel in question, e.g., for radiological AI diagnostics for radiology technicians and radiologists.

However, this consideration ought to be pushed even further if we assume that AI may become ubiquitous within clinical decision-making. On the one side, even if the final decision lies with a human expert, their decision may be ever more informed by algorithmic evidence that is unexplainable. The interpretability requirement, then, becomes ever more important: not only must the medical professional be able to interpret the machine's suggestions and decisional pathways, the professional must also be able to relate this information to the patient. Their expertise, thus, of those experts that does not only lie with medical knowledge anymore, which potentially can be produced more reliably by machines, but also with the apt communication of this information to the patient. While the trust relationship between such a medical professional and their patient ought to sustain the possible gaps of explanation, we ought to be aware that these gaps will emerge. That means that the expertise of medical professionals must be adequate to the tasks ahead, to create the trust necessary between them and their patients.

Medical informatics professionals, e.g., may become a new position within the medical field whose task it is to interpret and relate medical information produced by unexplainable technologies to patients. As experts in the prowess (but also the limits) of these machines, they can explain on higher levels what the decisional pathways amount to.

This would sufficiently fulfill the claim to expertise that every patient has. It does not, however, solve the injustices of our global healthcare system in which expertise is unequally and unjustly distributed (Freyer & Kempt, 2023).

5.2   Autonomous Cars and Weapons

5.2.1   Autonomous Cars

Autonomous cars are a contentious subject of AI use for a myriad of reasons (Cunneen et al., 2019; Nyholm, 2018). Next to concerns of the real-life trolley problem, coordination issues with human drivers, loss of autonomy in transportation, market domination, reliability in different weather conditions, etc., lies also one about the explainability of such systems. These cannot be resolved by trust, as, in opposite to medical devices, cars are not embedded into an established network of norms of expertise. While driver's licenses guarantee a minimum of expertise in operating a car, it is far from what is required to interpret a machine's behavior in a traffic situation.

Their explainability becomes a relevant factor in their justified use, similar to the medical domain, due to the higher stakes in operating these machines (Atakishiyev et al., 2024, Levinson et al., 2011). They carry the risk of seriously harming, if not killing both the user (i.e., the passengers) and/or other traffic participants, including pedestrians and bystanders (who are not, at that time, traffic participants, like guests of a restaurant with street-side seating).

The potential to harm others in the use of this technology sets up a different ethical debate: the decision subjects of the AI's decision are not only the users of that technology, and the decisions are not deliberately suggested with the recourse for contests, corrections, or other interventions, but potentially split-second decisions without any recourse. While the main passenger (i.e., the one in the driver's seat) currently holds some ability to intervene, and thus holds large parts of the blame if there are crashes and malfunctions caused by the self-driving car, neither other passengers nor other traffic participants and bystanders have an ability to intervene. This makes self-driving cars a prime example for responsibility gaps: How much can a driver rely on the AI's functions to not be able to intervene, and thus to create a responsibility gap in which the engineers as well as the driver are no longer suitable subjects of responsibility? The question of explainability can be best understood from that perspective and its' bearing on clarifying the responsibilities in self-driving cars, as well as closing the accountability and liability gaps that result from a responsibility gap.

Answering the question of who is responsible (or ultimately legally liable) for a crash caused by, e.g., a bad turn of a self-driving car, requires the opening of the decision-making process of the algorithm to reconstruct the timeline and to assess whether it was reasonable to expect the person in the driver's seat to react in time to avoid the crash. This reconstruction of the functioning of the car, especially in the pathways that led to the error, is crucial for establishing the causal chain we require to fulfill the minimum legal standard of liability. If we cannot establish such a causal chain (or an approximation thereof), we may not allow self-driving cars to drive in the first place.

However, the difference between the medical cases and the ones of self-driving cars is temporal. In medical cases, patients are asked to consent to their treatment after being informed by the decision-making pathways (or the approximation thereof). In self-driving cars, we aim to create reliable and yet "explainable" cars, so we can, if they fail, reconstruct the chain of causation to establish moral responsibility and legal liability. The question here, though, is to what degree explainability is required to fulfill this purpose. As we have seen, post-hoc rationalizability can be used to retro-fit the output to previous inputs. And since we have the data gathered by the self-driving car, we may be able to find the explanation when a certain course of action was decided, when that decision was communicated to the "driver", and how much time or opportunity the driver had to inter-vene. This all requires a human-in-the-loop as a "liability shield", on the one side, and a certain amount of transparency (see Sect. 2.5), on the other side. Whether this will suffice for sufficient certainty in legal consid-erations remains to be seen. Post-hoc-rationalizability may not, however, deliver the justification for full self-driving, as the ethical and legal issues emerging from fully autonomous, non-intervenable machines are beyond mere the explanation of an error after the fact. However, higher standards of explainability may help here only insofar as we will be able to democrat-ically decide (see Sect. 4.7) on the risks and levels of failure we tolerate, and the requirements for maintenance, safety updates, and cybersecurity to avoid these machines from being hackable.

Auditability (Sect. 2.8) might be suited best as a normative require-ment for these purposes. Since it allows independent controls for the proper function of these machines, and the necessary democratic control to avoid the control of those systems by adversarial powers, we can enable proper investigations for determining legal accountability. From the perspective of unexplainable technology, then, even fully autonomous cars may be justifiable if their key decision-making pathways are auditable, even when they are not explainable in any more granular sense. We may also consider some interpretability measures for passengers, as this might both improve the trust in those machines as well as the subjective comfort of "understanding" what the self-driving car is "thinking" (i.e., visualizations of the relevant elements in the car's decision).

5.2.2   Autonomous Weapons Systems (AWS)

In the discourse surrounding autonomous weapons systems, also known as "killer drones", explainability is usually not an aspect that is given much attention (see Wood's (2024) for why explainability has received little to no attention in military ethics and will likely remain so). This can be explained by the importance of other ethical questions surrounding the development and deployment of machines that kill based on their own decision-making structures: closing the responsibility gap and improving the transparency of their use are chief concerns, next to the changing character of warfare, the dehumanizing way of being killed, the desensitization of soldiers when their "dirty work" is ultimately nothing more than supervising and confirming kills, and the potential to be more surgical and life-saving if no actual combat takes place.

The key issue here for accounts that make explainability a normatively weighty concern is the fact that the decision subject can not only not agree to the treatment by an AWS but probably disagrees with the decision in the first place (as the decision is to be killed). That means that even though other humans are deploying AWS, it is the AWS that determines the target and attempts at eliminating it, and thus becomes the appropriate target for contestation of the decision.

Explainability of technology has thus far been sketched as an issue between the technology and the decision subject, as the institution that deploys that technology is not being interrogated for the appropriateness of their reasons. If that were the case, then the technology in question is the one to which someone would have a right to explanation. However, as it is in the nature of AWS, the decision subject of an AWS is usually unfortunate enough to not be able to contest this decision on principle grounds. Especially in cases of failure of the system, i.e., when it erroneously attacks civilians (as we have seen in the deployment of AWS in current conflicts), is not only the lack of accuracy in the decisional pathway morally concerning but also the irreversibility of the damages done. One way of ameliorating this issue is by validating the decisional pathways with higher restrictions, or by relegating the ultimate decision back to a human soldier. Neither appear to be promising avenues. Ultimately, as the decision of an AWS is about the life or death of the decision subject, the only way for explainability concerns to be addressed from the perspective of the decision subject is political. That would mean that based on the fact that we all might become decision subjects of an AWS which would violate our right to explanation, we should address their unexplainability before they are deployed.

However, attempting to enforce a right to explanation or any other explainability demand to the deployer of the weapon system changes the character of that charge: it is no longer about the explainability of a specific instance since that is practically impossible, but the transparency of the implementation and policy of using drones in warfare. It might be a weak argument in the complete picture of the ethical permissibility of AWS, but it provides a clear answer as to why being killed by a drone that made the determination to kill is a moral rights violation.

Further, the reason why explainability is usually not discussed as a separate issue but in association with responsibility concerns lies with questions surrounding the knowledge condition of responsibility. For anyone to reasonably take responsibility for a drone attack, even if the determination was done by the AI autonomously, is the knowledge gathered about how the drones make these determinations. This could, as mentioned above, be achieved by high validation requirements and some reliable interpretable systems, which can predict the behavior of a more sophisticated AI (see Sect. 2.3).

The risk here lies within the unlikely cooperation of militaries around the world, as higher demands for predictability and validation of machines used in combat take away their strategic advantage over the enemy, opening the door for cybersecurity exploitations (which is, noticeably, inverse to the unexplainability of self-driving cars: while their explainability is in the interest of the public to make them more robust against adversarial attacks, AWS become more vulnerable the more is publicly known about their construction). Auditability, in this sense, only limits the use of AWS rather than improving it.

Any military has good reason to keep some functioning technology a black box as long as instrumental reasons for strong explainability knowledge do not outweigh the "built-in secrecy". It makes AI resistant to being reverse-engineered, since a lack of knowledge about the construction of the AI keeps adversaries from retracing.

Ultimately, one way to address the issue of unexplainable AWS could be a political demand for very strong validation demands and a prohibition for anything that is below the required levels, even if that means that AWS becomes indefensible in use altogether. As we have seen, however, the lack of contestability or any other way for the decision subject to intervening with the decision provides a (however weak) argument against these systems and their deployment altogether.

5.3   Recommender Algorithms

Recommender algorithms are those algorithms that are able to provide suggestions (recommendations) of digital content to a user based on their past content consumption or other behavior (e.g., search queries in search engines, shopping websites, etc.). These recommender algorithms are virtually everywhere, as they have become both a necessity to navigate the vast amounts of content available and the difficulty to browse that content without proper guidance, and an often appreciated feature to be shown content that is relevant for a user's interests, preferences, or purposes.

This technology, thus, has an often underestimated influence on the digital lives of many, in many different forms. From economic influences (search engine and large shopping website rankings can make substantial differences in the sale of a product), to social and political ones (affective polarization, radicalization, and other political effects due to recommender algorithms on social media websites like Facebook and YouTube), the effects have become visible in shifting attitudes and profit margins.

Most—if not all—of these recommender algorithms exist within the private data ecosystem: the gathering of user-specific information (i.e., via cookies and other data-gathering means), the development of algorithms, their sale to other companies, and the deployment on websites all exist within the private market of digital data. This creates a specific set of incentive structures that are sought by those companies: some aim at gaining more data from the user in order to sell that data further, so others may create targeted ads; some aim at keeping the user on their platform for as long as possible to increase the amounts of advertisement; some aim at recommending user-specific products on their website to increase the amount of money spent on their website. The overall goal, as with most private enterprises, is to eventually create a profitable business: this means that at some point, the user's data becomes a product in itself, or the user will be compelled to become a customer. On websites that aim at selling ads (media- or social-media websites), recommender algorithms ensure that users remain on the site to be exposed to more (relevant) ads; on websites that sell products, like shopping or airfare sites, they aim at showing you products and prices that will make the user more likely to pay.

All these aim at engaging the user in ways that keep the user on websites that are essential parts of most users' everyday digital experiences. While users do usually not experience harm from these means, the nudges and other manipulative means to keep the user there may compound to harm from disorientation, autonomy-reduction, alienation, and severe misinformation. At the same time, continued, "tailor-made" misinformation and desensitization due to the presentation of highly divisive, offensive, politically extremist content can lead to a radicalization of users.

There is another argument to be had that mirrors the re-enchantment-concern from Sect. 4.1.2: the hubris present in some of the advertising and dealings with data from those companies (take, e.g., a music streaming service's claim to know their users' taste in music better than themselves) suggests that their decision-making AI is so sophisticated that it exceeds our own self-knowledge. Getting too good at predicting user behavior can have some creepy and arrogant dimensions, which might constitute a demand for thorough transparency and some interpretability measures. The fact that data can be used to fairly precisely predict people's political preferences, sexual orientation, or consumer habits thus is to a degree touching upon transparency questions.

Another instance of recommender algorithms that carry considerable moral weight is those used in formal, administrative decision-making. These are not customer-facing but support automated decision-making for specific purposes that are usually operating with statistical methods already (i.e., those regulated by the AI Act). Most prominently, these algorithms are used in deciding about loan, parole, or job applications, but can be used in many other kinds of ways where a (quasi-)formal decision-making complex is sought to be automated (i.e., life insurance premium calculations, claims to social security benefits, tax auditing process, etc.).

The principles of decision-making of these algorithms are similar to other recommenders: they combine personal and social statistical patterns of behavior to predict future behavior. In that sense, whether a website suggests you another shoe to buy that fits your (and other people's) shopping-behavior, or an algorithm recommends your application to a hiring committee, the principle remains the same. However, they are matching those behavior predictions with specific requirements from the decision-making institution: the decisions made about loan, parole, or job applications are based on results that one has to match in comparison to their expectations. A statistical likelihood of, say, 15% reoffending within the same year might be enough to disqualify someone from being granted parole, even though that particular case ensures that the person will not re-offend at all.

This makes the relevance of other people's behavior more important, relative to the decision than in other recommender algorithms. This is because those decisions are fairly rare to make by the decision subject (one does not regularly apply to loans or parole) so a personal behavioral pattern cannot be detected and factored in (only by indirect means). This can lead to unfair decision-making processes if a decision subject happens to carry a feature that the algorithms determined to be a useful predictor for future behavior.

This also introduces the concern of biased decisions caused by biased data. This well-researched issue demonstrates the vulnerability of automating certain socially relevant decisions. The use of training data, thus, is not only about the performance but also about the role of algorithmic decision-making: making decisions solely based on past decisions without assessing the merits of a case condemns us to stagnation. The correlation between certain demographic features, such as race, age, sex, or place of birth, and future behavior (defaulting on loans, reoffending after parole, job performance) can in a fully automated perspective lead often enough to seemingly clear decisions. These decisions, however, fully reduce each decision case to its statistical features, without reflecting on the reasons why these statistical features are of a certain kind, and whether they are fair in the first place: those applying algorithms to automate decisions with strong human dimensions may only care about the success-rate and other quantitative features; not about the moral quality of wronging people based on their demographic features or the severity of the individual consequences of being denied a loan, parole, or job despite qualifying.

Making these algorithms more explainable, thus, gains relevance as an issue for the safety of the user's mental health and autonomy (i.e., in being negatively affected mentally by recommended contents), perpetuation and deepening of bias and biased decisions through biased training sets, and the political stability (i.e., in being radicalized or misinformed by recommended contents) of their ubiquitous use. Accordingly, the ethical debate surrounding recommender algorithms has included explainability and associated considerations (cf. Zhang & Chen, 2020 for an overview). Of chief importance, however, appears to be transparency (Sect. 2.8). As we have seen in the chapter on transparency, making an algorithm more transparent usually does not require making the technology itself more understandable, but introduces some requirements for the contexts of creating the algorithm, deploying it, and the consequences of its deployment.

In the case of recommender algorithms, this means that there are, first, requirements for the transparency of the data-gathering process. The European Union's GDPR regulates part of this process, by making the use of cookies for advertisement purposes transparent: giving website users the choice of which cookies can be gathered makes this process clear. Whether or not this enables users to be in charge of their own data remains questionable, as a certain overwhelming impression can occur when browsing the internet, as every single website asks for one's own cookie policy; however, it has a limiting effect on the sale of data for targeted advertising.

Second, it requires transparency about the contents of the training data, which may require extensive testing, correction, and maintenance. This is to avoid the perpetuation and reproduction, if not even deepening, of biased and unfair decision-making procedures currently widespread in many areas of the world. If these are not corrected, then algorithms become a means of suppression and political control, as the implementation of seemingly impartial but ultimately opaque algorithms can be intended to keep this unfair bias alive.

Third, it means that the deployment of recommender algorithms may be monitored and constantly validated to avoid the inherently non-democratic, private interest of ever more engaging but politically harmful content taking over. Transparency requirements, here, can compel recommender algorithm using websites to provide reports about their use and potentially re-align them to the democratic goals prescribed by the lawmaking institutions.

Some cases of recommender algorithms are the target of those arguing for a right to explanation, especially to enforce the democratic control over private companies and the individual control over opaque collective decision-making processes. Especially when non-voluntary hierarchical systems are involved, being subjected to a decision by an algorithm that uses other people's pattern of behavior to deny one fair treatment appears to be a strong argument in favor of such right. Transparency requirements, additionally, will not be sufficient to satisfy the right even though they also are highly relevant for exerting democratic control over private entities. The mere transparency of which data has been used, however, will not provide the explanation for specific decisions. In these cases, we might expect the explanation to include the specific features that caused the decision: whether they played a role, and if yes, how weighty, they were. This could be covered by interpretable models, as they provide those with the final decision with sufficient insight into the algorithm and the decision behind it. Similar to medical AI, a certain certifiability-process could also determine the levels of risk acceptable to a society.

The issue, thus, is not that a single user has a right to know where a certain recommendation came from technologically, but that a user should be able to control whether or not this recommendation was based on one's own data, and, if the decision is of a certain relevance, that an expert will relate the decision to them. Most single instances of recommender algorithm are, from the perspective of ethics of technology, especially of explainability perspectives, rather uninteresting but representative of already problematic decision-making practices that miss the merits of an individual's case in favor of statistics-based decision-making. Thus, transparency and a serious democratic approach to these transparency requirements through a certification-process can govern recommender algorithms in a way that can weaken the need to explain the technology to any deeper depths.

5.4   Generative AI and General-Purpose Chatbots

Generative AI has become the most effective type of artificial intelligence as its progress over the last few years has reached both substantial levels of utility in text, image, audio, and video creation, as well as an economic and media hype with great promises to improve it even further. Multi-modal models have made it possible to create text-to-image, image-to-video, video-to-text, etc., combinations that essentially allow a generative model to describe and produce any kind of medium based on any kind of input. Their utility, despite the lack of reliability and safety, has been proven beyond doubt: voice-recognition and machine translation software has gotten so good compared to just a few years ago that research paradigms have shifted; chatbots are reliable enough that their use in many customer-service instances has become an attractive economic business model; and their human-imitation has become so convincing that we ought to seriously consider and carve out a space for their future in society.

With generative AI being so economically important that they will not go away, and yet so large that they have become unexplainable, ethical discourse has shifted towards their transparency. This includes, as previously discussed, the sources and uses of training data, the business model associated with training the models, and human labor and some sustainability assessments. All of these points have received scrutiny over the last few months and years as the industry of generative AI has exhibited weaknesses. First, their viability is questionable if we take copyright seriously. It appears that most generative models, both text- and image-generation, have used a lot of copyrighted material as training data that can, under some circumstances, be retrieved from that training data. And while generative AI-companies claim that this is both unavoidable (as otherwise it's prohibitively expensive to train the models and satisfy their requirements for data) and irrelevant (as the training data is merely used to *train* the model and not intended to ever feature in the models specifically), the retrieval of training data, or some overfitting-phenomena have documented that generative AI might always be vulnerable to reproducing some of its training data.

Second, annotating the training data and reinforcement learning with human feedback (RLHF) both have a high toll of human labor that often goes unnoticed in the wider debate. The training process is not only computationally demanding, but also labor-intensive. The amount of annotation in training data needed to make it usable for the large models requires human judgment. Further, the fine-tuning through human feedback requires human laborers to assess and correct the outputs of general-purpose chatbots. They are routinely confronted with very strong, offensive language as they have to give the feedback that this kind of language is inappropriate. The mental toll this kind of work takes is currently poorly understood, but ought to be acknowledged in the transparency of chatbots that have been trained with RLHF.

Third, both the training process and their continued use require large amounts of electricity and other energy, as well as freshwater as coolant, and other natural resources to build the GPU units. The environmental impact of generative AI especially has been appreciating in public interest, but still has been brushed aside by many of the large companies. Making the environmental footprint visible to AI and its training could help make the cost of AI more transparent.

Transparency for generative AI, thus, is an important step in making them more acceptable. However, in the current debate, questions of explainability of generative models are barely discussed. This might lie with the fact that (a) these models have become so large that any substantial explainability is unlikely, and (b) these models have become economically so important that their use is a given.

Is there Explainability to be Had?

If we were able to explain the decision-making pathways or the token-prediction to a degree that would allow for corrections or interventions (rather than statistical approximations), issues such as AI safety would be much less of a headache as it is as current stage: the instrumental benefit of predicability ensures a quicker intervention in safety-issues regarding their output, while it may also enable increased maintainability and the reduction of tech-debt as a result of widespread use of generative AI despite its flaws.

However, the latest developments in generative AI render any ambitious explainability virtually impossible. Especially few-shot/one-shot/zero-shot learning capabilities (i.e., methods within deep-learning that enable algorithms to predict new classes of subjects that were barely or not present in the training data (Chen et al., 2021)) reduces the likelihood to predict outputs, as there are barely any training-patterns to trace. These few-shot technological developments are very useful for areas in which only few or poor data is available, i.e., areas in which machine-learning algorithms previously had a hard time performing well. The ability to generalize from rich data and transfer those generalizations to areas with poor data complicates the explainability of algorithms even more, as one cannot even draw conclusions purely based on the training data and similar knowledge, i.e., predictors that are still explainable.

This, and some other indicators, have been claimed to be signs of emerging capabilities, i.e., previously unknown capabilities of a model that emerge during scaling up in model size, computational power, or the amount and quality of training data. If it was the case that scaling may lead to previously unknown and unpredictable skills, the epistemic explainability of LLMs would ultimately become an ontological one, as there may be no way of ever gaplessly explain the decision-making pathways. As we have seen in Sect. 2.2, ontological explainability renders the explanandum fundamentally unexplainable. Emerging capabilities, that cannot be predicted before scaling, might fall under this label.

Lastly, the phenomenology of interacting with these machines is also complicating any explanation that is based on explainability efforts as currently pursued: in many instances, LLMs appear to provide substantial reasons for their responses, and pretend that they reasoned through these reasons to get to the response that they provided. While the technological paradigm in LLMs does not seem to support these self-made claims by LLMs, the explainability of these machines is becoming ever more muddy through the technology itself providing assessments of its capabilities. A technology that provides reasons for its behavior would not merely *behave* anymore, but act, at least to some functional concept of agency. Whether we have reason to believe that an AI is at some point genuinely able to report its internal process adequately remains to be seen. However, once it does, the question of explainability does change from a mere causal explanation to then also include normative dimensions (as we have seen with Jongepier and Keymolen's characterization of a "useful explanation", see Sect. 4.4). This is a unique feature of general-purpose chatbots and other text-producing generative AI, as they can self-reference and thus "explain" their behavior. Here it becomes clear that most explainability-endeavors have practical reasons in mind that are associated with the attribution of responsibility in case of failure (as with self-driving cars, AWS), or for being the guarantor of the entire process (as with medical processes). Mere self-explanation will not do, even if the explanations provided are somewhat plausible and sound like something a moral agent would do.

The issue, then, is that blameworthiness and responsibility are only applicable to moral agents, while the ability to explain one's reasons for behaving (i.e., acting) is not necessarily associated with moral agency: as Floridi puts it, LLMs are agents without intelligence (Floridi, 2023), and thus seem to behave like agents but do not perform tasks the same way agents do: their self-reported intentions and reasons are not, in fact, the reason why they do anything. These self-reported reasons are merely the approximation of what an agent would say if they were asked for their reasons.

As we have seen above with recommender algorithms, the damage done in cases of malfunction or errors to individual users is not as great as it is the case in autonomous cars, AWS, or medical diagnostics. The concern is, again, the compounding effect of thousands of problematic utterances. However, in opposite to recommender algorithms, generative AI does hold the potential to cause harm to individual users by providing false information, psychological manipulation, offensive language, etc. A morally reliable solution that can be technically implemented is still missing, however, and the discourse on what actually should be considered "safe" is still undecided and rather confused (see Kempt, Lavie, & Nagel (in press) for a proposed way to assess the safety of LLM-utterances via the concept of appropriateness).

5.5   How to Explain Unexplainable Technology

The fact that we have to deal with technology that eludes our ability to offer causal explanations for its behavior has been a controversial subject of debate for those regulating and implementing those technologies. The examples from this chapter demonstrate that the mere fact that an algorithm may be not fully explainable does not determine the normative demand for it to be made (more) explainable in order to be justifiably used. Three elements have been shown to alleviate these concerns:

First, a highly accurate and reliable algorithm, to which there are only costly alternatives, may be deemed to be outweighing the normative cost of it being unexplainable.

Depending on the point of comparison (i.e., other machines that are explainable, human labor, etc.), the requirements for accuracy and reliability of a machine's output can outweigh the lack of understanding of why some of the machine's behaviors produce such accurate and reliable results. Understanding that the permissibility of a given technology ought to depend on its comparative qualities to other options, rather than its own features absolute of other considerations, is the key normative insight to deploying unexplainable technology. For most purposes, explainability is an instrumental reason that improves other features we are interested in, such as maintainability, error-shooting, or reliability overall. As authors like Alex London have pointed out (Ghassemi et al., 2021, London, 2020), explainability is often secondary if the accuracy is sufficiently high. Especially in areas in which human decision-making is subject to high variation of skill and bias can AI establish reliability that outweighs explainability requirements.

Second, the importance of a machine with a reasonable accuracy in an important function that has no point of comparison to human capabilities. The standards of explaining a process are usually laid on human experts who either perform or supervise that process. These standards stem from moral norms that are a result of recognizing decision subjects' claims, the ability and feasibility of explaining certain processes, some traditions of recognizing expertise, and the normative relevance of the process itself. If a machine performs an action that was previously not performed by a human being, but that is nonetheless important to human society, the question of what an explainability standard ought to be set is less salient than in a situation where a machine is taking over or performing similar tasks to human experts. Thus, for explainability discussions to have normative relevance, we ought to first determine if, and if yes, to what degree, "automation" is being performed here.

Third, the lack of stakes in the successful performance of the application. The assumption often seems to be that because (unexplainable) artificial intelligence is involved that the task performed must be of a certain level of cognitive complexity and thus non-trivial. However, many tasks AI is automating, partially even explicitly designed to do so, are low-stakes, high-effort tasks, or simply minor decisions. The single instances of an AI's behavior, especially those of generative AI and of recommender algorithms, are often of virtually no moral relevance. This does not mean that their entire activity is morally irrelevant: there are levels of description of what an AI is doing that are, in fact, morally relevant. The decision of a self-driving car, for example, to accelerate at a certain speed without anyone else present on the road might not be morally relevant as well, but the action of "navigating traffic" surely is. Both are "done" by the AI, in this sense. However, some full description of an AI's task are not reaching moral relevance. Take, in opposite to self-driving cars, a recommender algorithm on a shopping website for clothing. There is little else that the algorithm does but find patterns in both clothing style and buying patterns of customers and tries to find additional/alternative clothes that both have a high chance of being bought by the customer, or similar items that have a higher profit if bought instead. Neither reaches morally relevant stakes, and thus, its explainability is of little relevance as well.

We might also want to consider large parts of generative AI to fall in the categories 2 and 3, depending on their use: most of what generative AI like chatbots are doing is of little moral relevance as a uniquely automated affair. This does not mean that the use is frivolous or entirely pointless. Rather, that the moral relevance of it being done by an AI is not bearing on the question if we fully understand how the machine works.

Take, for example, the automated formulation and submission of spam to social media, search engines, and the like. The so-called enshittification (Birch, 2023) of the internet by generative AI is going to be a long-term problem of keeping the internet as a realm that can productively be navigated by humans (rather than, say, with the help of yet another AI to avoid these purely AI-generated exchanges, news sources, etc. that have little to no connection to actual human contributions). It is, however, not a fully new phenomenon, but merely an acceleration of a trend that has been noticeable over the last decade. Thus, explainability as a question to put forward about the technology used is less of interest than, say, questions of responsibility for releasing these algorithms, the safety-measures implemented to suppress the ability to spam the internet, or the detectability of their utterances. Reasons for pursuing explainability-questions do not lie alone within the technology itself, but also in the practical circumstances of its implementation.

However, we can conclude that the stakes of the decisions made by an AI do bear on the requirement for explainability at least to the degree that the utility of explainability improves.

References

Amann, J., Blasimme, A., Vayena, E. et al. (2020). Explainability for artificial intelligence in healthcare: A multidisciplinary perspective. *BMC Med Inform Decis Mak, 20*(310). https://doi.org/10.1186/s12911-020-01332-6

Atakishiyev, S., Salameh, M., Yao, H., & Goebel, R. (2021). *Explainable artificial intelligence for autonomous driving: A comprehensive overview and field guide for future research directions.* arXiv preprint arXiv:2112.11561

Beauchamp, T. L., & Childress, J. F. (2001). *Principles of biomedical ethics* (5th ed.). Oxford University Press.

Birch, K. (2023). Data paradoxes. In *Data enclaves*. Palgrave Macmillan. https://doi.org/10.1007/978-3-031-46402-7_6

Bjerring, J. C., & Busch, J. (2021). Artificial intelligence and patient-centered decision-making. *Philosophy & Technology, 34*(2), 349–371. https://doi.org/10.1007/s13347-019-00391-6

Braun, M., Hummel, P., Beck, S., & Dabrock, P. (2020). Primer on an ethics of AI-based decision support systems in the clinic. *Journal of Medical Ethics Published Online, 47*(12). https://doi.org/10.1136/medethics-2019-105860

Chen, J., Geng, Y., Chen, Z., Horrocks, I., Pan, J. Z., & Chen, H. (2021). *Knowledge-aware zero-shot learning: Survey and perspective.* arXiv preprint arXiv:2103.00070

Cunneen, M., Mullins, M., & Murphy, F. (2019). Autonomous vehicles and embedded artificial intelligence: The challenges of framing machine driving decisions. *Applied Artificial Intelligence, 33*(8), 706–731. https://doi.org/10.1080/08839514.2019.1600301

Floridi, L. (2023). AI as agency without intelligence: On ChatGPT, large language models, and other generative models. *Philosophy and Technology, 36*(1).

Freyer, N., & Kempt, H. (2023). AI-DSS in healthcare and their power over health-insecure collectives. In H. Bhakuni & L. Miotto (Eds.), *Justice in Global Health* (pp. 38–55). Routledge.

Ghassemi, M., Oakden-Rayner, L., & Beam, A. L. (2021). The false hope of current approaches to explainable artificial intelligence in health care. *Lancet Digit Health, 11*, e745–e750. https://doi.org/10.1016/S2589-7500(21)00208-9

Grote, T., & Berens, P. (2020). On the ethics of algorithmic decision-making in healthcare. *Journal of Medical Ethics, 46*(3), 205–211. https://doi.org/10.1136/medethics-2019-105586

Kempt, H., Freyer, N., & Nagel, S. K. (2022). Justice and the normative standards of explainability in healthcare. *Philosophy and Technology, 35*(100). https://doi.org/10.1007/s13347-022-00598-0

Kempt, H., Heilinger, J. C., & Nagel, S. K. (2022). Relative explainability and double standards in medical decision-making. *Ethics and Information Technology, 24*(20). https://doi.org/10.1007/s10676-022-09646-x

Kundu, S. (2021). AI in medicine must be explainable. *Nat Med 27*(8), 1328. https://doi.org/10.1038/s41591-021-01461-z

Levinson, J., Askeland, J., Becker, J., Dolson, J., Held, D., Kammel, S., Kolter, J. Z., Langer, D., Pink, O., Pratt, V., Sokolsky, M., Stanek, G., Stavens, D., Teichman, A., Werling, M., & Thrun, S. (2011). Towards fully autonomous driving: Systems and algorithms. In *2011 IEEE Intelligent Vehicles Symposium (IV)* (pp. 163–168). IEEE.

London, A. J. (2019). Artificial intelligence and black-box medical decisions: Accuracy versus explainability. *Hastings Center Report, 49*(1), 15–21. https://doi.org/10.1002/hast.973

Nyholm, S. (2018). The ethics of crashes with self-driving cars: A roadmap. *Philosophy Compass.* https://compass.onlinelibrary.wiley.com/doi/full/10.1111/phc3.12507 (last accessed May 31st 2024)

Price II, W. N. (2015). Black-Box medicine. *Harvard Journal of Law & Technology, 28*(2).

Rudin, C. (2019). Stop explaining black box machine learning models for high stakes decisions and use interpretable models instead. *Nature Machine Intelligence., 1*(5), 206–215.

Rueda, J., Rodríguez, J. D., Jounou, I. P., Hortal-Carmona, J., Ausín, T., & Rodríguez-Arias, D. (2022). 'Just' accuracy? Procedural fairness demands explainability in AI-based medical resource allocations. In *AI & Society, 39*(3), 1–12.

Sand, M., Durán, J. M., & Jongsma, K. R. (2021). Responsibility beyond design: Physicians' requirements for ethical medical AI. *Bioethics*. https://doi.org/10.1111/bioe.12887

Theunissen, M., & Browning, J. (2022). Putting explainable AI in context: Institutional explanations for medical AI. *Ethics and Information Technology, 24*, 23. https://doi.org/10.1007/s10676-022-09649-8

Ursin, F., Timmermann, C., & Steger, F. (2022). Explicability of artificial intelligence in radiology: Is a fifth bioethical principle conceptually necessary? *Bioethics, 36*(2), 143–153. https://doi.org/10.1111/bioe.12918

Wadden, J. J. (2022). Defining the undefinable: The black box problem in healthcare artificial intelligence. *Journal of Medical Ethics, 48*, 764–768.

Wood, N. G. (2024). Explainable AI in the military domain. *Ethics and Information Technology, 26*, 29. https://doi.org/10.1007/s10676-024-09762-w

Zhang, Y., & Chen, X. (2020). Explainable recommendation: A survey and new perspectives. *Foundations and Trends in Information Retrieval, 14*(1), 1–101.

CHAPTER 6

Explainability Debunked?

Abstract Having elaborated on the epistemological conditions and the ethical interplay between explainability and expertise, we have found that explainability does not have a purpose beyond its instrumental value for other purposes. It might even be problematic to insist on explainability too rigorously. Since the applied cases of unexplained technology have demonstrated that explainability is often not the technique required for ethical permissibility, we can elaborate on whether explainability is practically relevant in the first place. Especially considering the emergence of ever-growing foundational models has undercut hopes to meaningfully increase explainability in AI.

Keywords Explainability · Foundational models · Justice · Transparency

We have many reasons for putting great efforts into increasing our ability to appreciate and understand the decisional pathways of technology that remain otherwise opaque to us. However, it is important to get the reasons right why we should care about the fact that some unexplainable technology is not only possible but also takes hold in digitizing societies. As we have seen, there are some clear intuitions that unexplainable technology should be explained. These intuitions, however, necessitate the discussion of some questions that emerge when arguing for explainability requirements: what is the subject of the explanation? To what degree of granularity should the technology be explained? How can we avoid picking too fine-grained or too rough-grained levels that do not deliver any understanding relevant to informed decision-making? The work undertaken in xAI is not always useful in answering these broader questions and does not have to be: as its own sub-discipline of research, it may pursue tangential, epistemic-purpose-driven goals rather than practical ones.

The core question for the ethics of explainability, then, is this: Just how important is explainability for most of our digital lives and purposes therein? We have found a potentially controversial, but I hope convincingly argued-for position that explainability has many benefits, but none of them preclude unexplainable technology to be used if explanations were not available to a minimal degree. If a technology is unreliable and we do not understand why, understanding why it is unreliable will not make it more likely to be used: rather, making technology more reliable is the key to acceptable use of it. The fact that we work with risk assessments suggests already that a lack of explainability can be overcome if the accuracy and validation are of high enough precision. Any level of explainability that helps with that is valuable because of its instrumental status in improving the statistics of the technology we do actually care about. The remaining normative concerns are covered by our expertise practices that have grown into reliable, controlled, and essential ecosystems within our ways of life that we cannot live without.

From this perspective, many of the explainability efforts seem like thinly veiled attempts at making a technology more accepted by making it seem more acceptable. However, explainability has been shown to be one of our lesser concerns in making AI acceptable: many applications do not suffer from a lack of explainability but from transparency. The reasons for using AI, not their specific implementations, are of high moral relevance in progressing towards a more acceptable AI-landscape.

As no technology will ever guarantee full safety, even with risk–benefit analyses with highly desirable results, we ought to reckon that humans will be harmed by AI decisions. Some technologies, like AWS, may even cause death if they are making the right call or the wrong one. Those who will be affected by the errors made will have to be compensated in a similar fashion as those currently affected by the mistakes of experts. This represents a legal challenge that has to be addressed by lawmakers, especially when considering that the use might not be only in private hands but in public ones, with no alternative or right to not be subjected to these systems' decisions. Does it, however, affect moral considerations that much?

Colaner (2022) claims that the lack of explainability constitutes a dehumanizing feature in AI decision-making that gives rise to considerations of outlawing it. However, he does not engage in is a weighing of benefits and drawbacks, i.e., of acknowledging the opportunity costs of the consequences of his proposal. If we were to agree that unexplainable technology is a violation of our human dignity, we might be inclined to reject AI-based decision-support systems altogether until we have figured out how to make them more explainable. Ultimately, we ought to keep in mind that we are dealing with a technology that has not matured yet, but already promises to improve the lives of many manifold. Limitations to that technology, thus, ought to be first and foremost motivated by justice and distributive considerations.

Explainability can only help in those areas by improving those things we actually care about. There might be some higher-level demands for understanding when an automated decision causes great harm to the decision subject (be it by misdiagnosing a patient, rejecting an otherwise quality parole applicant, or killing an innocent refugee in war), but most concerns with wrong decisions are about compensation and retribution. For that, explainability is of limited use.

6.1 THE STRUGGLE AHEAD WITH GENERATIVE AI AND FOUNDATIONAL MODELS

Considering the struggle with explainability in models from the late 2010s and early 2020s, we should not be surprised by the fact that in the following iterations of complexity and scale of neural networks, the ability to explain them in the initial sense appears hopeless. Ever-growing concerns about foundational models, i.e., those models behind highly skilled chatbots and image- and video-generating software, will not decrease through explainability demands.

We have seen that highly interactive and personal AI, such as chatbots and personal assistants, may provide explanations for certain behaviors as if they were reasons. This is both a concern and an illustration of the varying receptiveness of human beings to explanations. The intentional stance towards these machines may become non-optional in the future. The differences in explanations we expect and require to make our own decisions are reflected in this receptiveness and change of industry and technological sophistication. For some decisions, knowing that we were informed with the best available prediction about what will happen seems to suffice.

The larger concern from all kinds of generative AI in regard to their opacity appears to be the training data on which they were trained, the ability to retrieve (or reproduce) the training data, and the question of whether copyright applies to using data as mere training data, even if it not supposed to be retrievable. Current court cases and legal proceedings suggest that we eventually will settle on the question of the permissibility of using copyrighted data for the training of LLM, or not. If the latter is the case, we might see a considerable reduction on progress over the next few years in generative AI, as much of the progress is powered by speculation for financial returns that would certainly be diminished if generative AI companies would have to pay for copyrighted data (or even figure out which data sets contain copyrighted data). We might even witness the collapse of the entire industry under the burden of copyrighted material and training data requirements. Transparency, thus, is the key to understanding modern technology, because it is the key to understanding contemporary economic circumstances of generative AI: those who hold the data and the computing power hold the actual power.

Reference

Colaner, N. (2022). Is explainable artificial intelligence intrinsically valuable? *AI & Society, 37*, 231–238. https://doi.org/10.1007/s00146-021-01184-2

CHAPTER 7

Conclusions

Abstract By summarizing the points made in this book, this chapter concludes the line of argument. Explainability appears less relevant in most contexts, and in those in which it does, having expert advocates work on your behalf is the precedented, more established, and less disruptive solution.

Keywords Explainability · Democracy · Expertise · Right to Explanation

The fact that we have become able to create technology which we cannot explain to the degree we are used to is a philosophically wondrous fact and yet no cause for panic. In this book, I have attempted to summarize some of my research which is guided by the struggle to defend a very promising, soon-to-be (if not already) ubiquitous technology against the concern that it is lacking something that we usually consider important. The main points I tried to make in this book were the following.

1. Explainability plays an outsized role in philosophical discussions, especially when it is discussed as an ethical requirement in itself. This is understandable insofar as explainability often is a key part of a larger normative justification of otherwise unexplainable technologies: explainability is *useful* for making reliable predictions,

H. Kempt, *(Un)explainable Technology*, https://doi.org/10.1007/978-3-031-68098-4_7

maintaining the technology's accuracy and functioning, avoiding a re-enchantment of the world and deification of AI technologies through lacking AI literacy and fueled by corporate-driven AI-hype, and the like. Whenever explainability is important, however, it is mostly instrumentally important.

2. A lot of similar-looking but highly different AI are being used right now, let alone highly different but unexplainable technologies at large. The blanket claim that "unexplainable technology is lacking normatively important features and thus ought to be viewed skeptically" cannot be upheld, as well as its inverse statement: we cannot easily generalize the normative relevance of explainability. The huge variety of use of autonomous technologies for important decisions triggers different questions regarding their ethical permissibility: from sorting algorithms to generative AI, from autonomous weapons systems to medical diagnostics, they all come with different challenges in regard to their lack of explainability. AI is not a black box, but several different boxes with different shades of gray for which we require different levels of elucidation.

3. The right to explanation, as purported by some, is also not doing the normative grounding that is needed to outweigh the disruption this right will have on our relationship between explanations, expertise, and trust.

This is because expertise, as it currently functions in our high-trust societies, works differently. The widespread and deep silos of expertise guarantee a lifestyle and level of well-being that would otherwise not be possible. This, however, comes with the absolving of experts to explain their decision-making pathways to lay audiences, usually the decision subjects but also society at large. Adding artificial intelligence to this system is not to be understood as a normative disruption but as an addition of an ongoing process. Demanding of an AI-expert-composition to provide an explanation in the sense of a knowledge- or reason-producing one, can be overly costly in the general functioning of high-trust societies. Further, the great variety of how decision subjects need to be explained to achieve their subjective need for explanations creates an even more difficult translational issue from automated decision-making to explainability, essentially minimizing the benefits of the spread-out system of expertise.

And while contestability works as a means of questioning a decision made by an opaque decision-making system, this can be better translated into a claim to expertise, rather than a right to explanation. This way, the decision and the interpretation, or explanation, of an opaque system lies with an expert who can assess—but does not have to explain—whether the decision made is appropriate and proper.

4. The need to democratically control unexplainable technologies might require higher standards of explainability than any moral consideration could provide. The moral and political authority of democracies to enforce their standards on private companies and their products, especially in terms of safety, security, maintainability, and technological sovereignty is a key element in governing contemporary technological developments.

What this standard-setting authority ultimately should settle on is an open political question, however, and thus depends on the political preferences of the respective administrations, the electorate, and other rules that might bind or free companies. There are ways, I have argued, that strong transparency requirements for the automation of decisions should be implemented. The drawbacks of such strong transparency rules usually do not outweigh the ethical requirements and long-term benefits.

However, there are justice concerns emerging in legislating standards for explainability, auditability, contestability, transparency, etc., that could limit the export of highly useful, quality-of-life-improving machines to other places. Setting explainability and transparency standards in a highly expertise-saturated, individualized context in which both the claim to expertise and the right to explanation are strong normative arguments may hinder the technology's ability to lead to substantial improvements elsewhere in the world, where those standards are simply lower (cf. Kempt et al., 2022). These considerations ought to feature in ethical considerations about dimensions of justice in determining standards.

Explainability as a philosophical concern, legal issue, and technological challenge will remain central to questions surrounding AI and algorithmic decision-making. The human condition to need to understand specifics even beyond the usefulness of such specifics will make unexplainable technology a sore spot in every inquisitive mind. Even the unfailing oracle, ultimately, will be asked to provide an explanation for its predictions.

REFERENCE

Kempt, H., Freyer, N. & Nagel, S. K. (2022). Justice and the normative standards of explainability in healthcare. *Philosophy and Technology, 35*(100). https://doi.org/10.1007/s13347-022-00598-0

Bibliography

AI Act (2024). https://artificialintelligenceact.eu/de/ (last accessed May 31st 2024).

Amann, J., Blasimme, A., Vayena, E. et al. (2020). Explainability for artificial intelligence in healthcare: A multidisciplinary perspective. *BMC Med Inform Decis Mak*, *20*(310). https://doi.org/10.1186/s12911-020-01332-6

Armstrong, S., Sandberg, A., & Bostrom, N. (2012). Thinking inside the box: Controlling and using an oracle AI. *Minds & Machines, 22*(4), 299–324. https://doi.org/10.1007/s11023-012-9282-2

Atakishiyev, S., Salameh, M., Yao, H., & Goebel, R. (2021). *Explainable artificial intelligence for autonomous driving: A comprehensive overview and field guide for future research directions.* arXiv preprint arXiv:2112.11561

Beauchamp, T. L., & Childress, J. F. (2001). *Principles of biomedical ethics* (5th ed.). Oxford University Press.

Berry, D. M. (2021). Explanatory publics. Explainability and democratic thought. In B. Balaskas & C. Rito (Eds.), *Fabricating publics: The dissemination of culture in the post-truth era.* Open Humanities Press.

Birch, K. (2023). Data paradoxes. In *Data enclaves*. Palgrave Macmillan. https://doi.org/10.1007/978-3-031-46402-7_6

Bjerring, J. C., & Busch, J. (2021). Artificial intelligence and patient-centered decision-making. *Philosophy & Technology, 34*(2), 349–371. https://doi.org/10.1007/s13347-019-00391-6

Brajovic, D., Renner, N., Goebels, V. P., Wagner, P., Fresz, B., Biller, M., Klaeb, M., Kutz, J., Neuhüttler, J., & Huber, M. F. (2023). *Model reporting for certifiable ai: A proposal from merging eu regulation into ai development.* arXiv preprint arXiv:2307.11525

Braun, M., Hummel, P., Beck, S., & Dabrock, P. (2020). Primer on an ethics of AI-based decision support systems in the clinic. *Journal of Medical Ethics Published Online, 47*(12). https://doi.org/10.1136/medethics-2019-105860

Burrell, J. (2016). How the Machine 'thinks': Understanding opacity in machine learning algorithms. *Big Data & Society, 3*(1), 1–12.

Byrne, R. M. (2023). Good explanations in Explainable Artificial Intelligence (XAI): Evidence from human explanatory reasoning. In *Proceedings of the thirty-second international joint conference on artificial intelligence*. International Joint Conferences on Artificial Intelligence Organization, Macau, SAR China (pp. 6536–6544).

Castelvecchi, B. (2016). *Can we open the black-box of AI?* https://www.nature.com/news/can-we-open-the-black-box-of-ai-1.20731 (last accessed May 31st 2024).

Chazette, L., Brunotte, W., & Speith, T. (2021). Exploring explainability: A definition, a model, and a knowledge catalogue. In *IEEE 29th International Requirements Engineering Conference (RE)* (pp. 197–208). IEEE. https://doi.org/10.1109/RE51729.2021.00025

Chen, J., Geng, Y., Chen, Z., Horrocks, I., Pan, J. Z., & Chen, H. (2021). *Knowledge-aware zero-shot learning: Survey and perspective*. arXiv preprint arXiv:2103.00070

Colaner, N. (2022). Is explainable artificial intelligence intrinsically valuable? *AI & SOCIETY, 37*, 231–238. https://doi.org/10.1007/s00146-021-01184-2

Cunneen, M., Mullins, M., & Murphy, F. (2019). Autonomous vehicles and embedded artificial intelligence: The challenges of framing machine driving decisions. *Applied Artificial Intelligence, 33*(8), 706–731. https://doi.org/10.1080/08839514.2019.1600301

Danaher, J. (2016). The threat of algocracy: Reality, resistance and accommodation. *Philos. Technol., 29*, 245–268. https://doi.org/10.1007/s13347-015-0211-1

Danks, D. (2022). Governance via explainability. In Justin B. Bullock, Yu-Che Chen, Johannes Himmelreich, Valerie M. Hudson, Anton Korinek, Matthew M. Young & Baobao Zhang (Eds.), *The oxford handbook of AI governance*. Oxford University Press.

de Fine Licht, K., & de Fine Licht, J. (2020). Artificial intelligence, transparency, and public decision-making: Why explanations are key when trying to produce perceived legitimacy. *AI & Society, 1–10*. https://doi.org/10.1007/s00146-020-00960-w

Dennett, D. C. (1981). True believers : The intentional strategy and why it works. In A. F. Heath (Ed.), *Scientific explanation: Papers based on Herbert spencer lectures given in the University of Oxford* (pp. 150–167). Clarendon Press.

Doshi-Velez, F., & Kim, B. (2017). Towards a rigorous science of interpretable machine learning. *arXiv* preprint arXiv:1702.08608.

Durán, J. M. (2021). Dissecting scientific explanation in AI (sXAI): A case for medicine and healthcare. *In Artificial Intelligence, 297,* 103498. https://doi.org/10.1016/j.artint.2021.103498

Edwards, L., & Veale, M. (2017). Slave to the algorithm? Why a "right to explanation" Is probably not the remedy you are looking for. *Duke Law & Technology Review, 16,* 18–84. https://doi.org/10.2139/ssrn.2972855

Elsen, J., Cizer, O., & Snellings, R. (2013). Lessons from a lost technology: The secrets of Roman concrete. *American Mineralogist, 98*(11–12), 1917–1918.

Erasmus, A., Brunet, T. D. P., & Fisher, E. (2021). What is interpretability? *Philosophy & Technology, 34,* 833–862.

Floridi, L. (2023). AI as agency without intelligence: On ChatGPT, large language models, and other generative models. *Philosophy and Technology, 36*(1).

Floridi, L., & Cowls, J. (2019). A unified framework of five principles for AI in society. *Harvard Data Science Review, 1*(1). https://doi.org/10.1162/99608f92.8cd550d1

Freyer, N., & Kempt, H. (2023). AI-DSS in healthcare and their power over health-insecure collectives. In H. Bhakuni & L. Miotto (Eds.), *Justice in Global Health* (pp. 38–55). Routledge.

Fricker, E. (2021). Epistemic self-governance and trusting the word of others: Is there a conflict? In J. Matheson & K. Lougheed (Eds.), *Epistemic autonomy* (pp. 323–342). Routledge.

General Data Protection Regulation of the EU (GDPR) (2016). 679. https://gdpr-info.eu/

Ghassemi, M., Oakden-Rayner, L., & Beam, A. L. (2021). The false hope of current approaches to explainable artificial intelligence in health care. *Lancet Digit Health, 11,* e745–e750. https://doi.org/10.1016/S2589-7500(21)00208-9

Goldman, A. I. (2001). Experts: Which ones should you trust? In *Philosophy and Phenomenological Research, 63*(1), 85–110.

Goodman, B., & Flaxman, S. (2016). *European union regulations on algorithmic decision-making and a 'right to explanation'.* https://arxiv.org/abs/1606.08813

Grote, T., & Berens, P. (2020). On the ethics of algorithmic decision-making in healthcare. *Journal of Medical Ethics, 46*(3), 205–211. https://doi.org/10.1136/medethics-2019-105586

Guerrero, A. A. (2017). Living with ignorance in a world of experts. In Rik Peels (Ed.), *Perspectives on ignorance from moral and social philosophy* (pp. 156–185). Routledge.

Günther, M., & Kasirzadeh, A. (2022). Algorithmic and human decision making: For a double standard of transparency. *In AI & Society, 37*, 375–381.

Gurrapu, S., Kulkarni, A., Huang, L., Lourentzou, I., & Batarseh, F. A. (2023). Rationalization for explainable NLP: A survey. *Frontiers in Artificial Intelligence, 6*.

Grossmann, I., et al. (2023). AI and the transformation of social science research. *Science, 380*, 1108–1110.

Haugeland, J. (1985). *Artificial intelligence: The very idea*. MIT Press.

Herzog, C. (2022). On the risk of confusing interpretability with explicability. *AI Ethics, 2*, 219–225. https://doi.org/10.1007/s43681-021-00121-9

International Organization for Standardization (ISO) (2020). IEC TR 29119–11:2020.

Jongepier, F., & Keymolen, E. (2022). Explanation and agency: Exploring the normative- epistemic landscape of the "Right to explanation". *Ethics Inf Technol 24*, 49. https://doi.org/10.1007/s10676-022-09654-x

Kästner, L., Langer, M., Lazar, V., Schomäcker, A., Speith, T., & Sterz, S. (2021). On the relation of trust and explainability: Why to engineer for trustworthiness. In *IEEE 29th International Requirements Engineering Conference Workshops (REW)* (pp. 169–175). IEEE. https://doi.org/10.1109/REW 53955.2021.00031

Kawamleh, S. (2022). Against explainability requirements for ethical artificial intelligence in health care. *AI and Ethics., 29*, 1–6.

Kasirzadeh, A. (2021). *Reasons, values, stakeholders: A Philosophical framework for explainable artificial intelligence*. https://arxiv.org/pdf/2103.00752

Kempt, H., & Nagel, S. K. (2021). Responsibility, second opinions and peer-disagreement: Ethical and epistemological challenges of using AI in clinical diagnostic contexts. *Journal of Medical Ethics, 48*, 222–229. https://doi.org/10.1136/medethics-2021-107440

Kempt, H., Heilinger, J. C., & Nagel, S. K. (2022). Relative explainability and double standards in medical decision-making. *Ethics and Information Technology, 24*(20). https://doi.org/10.1007/s10676-022-09646-x

Kempt, H., Freyer, N., & Nagel, S. K. (2022). Justice and the normative standards of explainability in healthcare. *Philosophy and Technology, 35*(100). https://doi.org/10.1007/s13347-022-00598-0

Kempt, H., Lavie, A., & Nagel, S. K. (in press). *Towards a conversational ethics for LLM*. American Philosophical Quarterly.

Krishnan, M. (2020). Against interpretability: A critical examination of the interpretability problem in machine learning. *Philosophy & Technology, 33*, 487–502.

Kundu, S. (2021). AI in medicine must be explainable. *Nat Med 27*(8), 1328. https://doi.org/10.1038/s41591-021-01461-z

Landgrebe, J. (2022). Certifiable AI. *Applied Science, 12*(3), 1050. https://doi.org/10.3390/app12031050

Langer, M., Oster, D., Speith, T., Hermanns, H., Kästner, L., Schmidt, E., Sesing, A., & Baum, K. (2021). What do we want from explainable artificial intelligence (XAI)?—A stakeholder perspective on XAI and a conceptual model guiding interdisciplinary XAI Research. *Artificial Intelligence, 296.* https://doi.org/10.1016/j.artint.2021.103473

Le Ludec, C., Cornet, M., & Casilli, A. A. (2023). The problem with annotation. Human labour and outsourcing between France and Madagascar. *Big Data & Society, 10*(2). https://doi.org/10.1177/20539517231188723

Leichtmann, B., Humer, C., Hinterreiter, A., Streit, M., & Mara, M. (2023). Effects of explainable artificial intelligence on trust and human behavior in a high-risk decision task. *Computer in Human Behavior, 39*, 107539. https://doi.org/10.1016/j.chb.2022.107539

Levinson, J., Askeland, J., Becker, J., Dolson, J., Held, D., Kammel, S., Kolter, J. Z., Langer, D., Pink, O., Pratt, V., Sokolsky, M., Stanek, G., Stavens, D., Teichman, A., Werling, M., & Thrun, S. (2011). Towards fully autonomous driving: Systems and algorithms. In *2011 IEEE Intelligent Vehicles Symposium (IV)* (pp. 163–168). IEEE.

Lipton, Z. C. (2018). The mythos of model interpretability: In machine learning, the concept of interpretability is both important and slippery. *Queue, 16*(3), 31–57. https://dl.acm.org/doi/pdf/10.1145/3236386.3241340

Lombrozo, T. (2011). The instrumental value of explanations. *Philosophy Compass, 6*(8), 539–551. https://doi.org/10.1111/j.1747-9991.2011.00413.x

London, A. J. (2019). Artificial intelligence and black-box medical decisions: Accuracy versus explainability. *Hastings Center Report, 49*(1), 15–21. https://doi.org/10.1002/hast.973

Maclure, J. (2021). AI, explainability and public reason: The argument from the limitations of the human mind. *Minds and Machines, 31*(2021), 421–438.

Maung, H. H. (2017). The causal explanatory functions of medical diagnoses. *Theoretical Medicine and Bioethics, 38*(1), 41–59. https://doi.org/10.1007/s11017-016-9377-5

Messeri, L., & Crockett, M. J. (2024). Artificial intelligence and illusions of understanding in scientific research. *Nature, 627*, 49–58. https://doi.org/10.1038/s41586-024-07146-0

Miller, T. (2019). Explanation in artificial intelligence: Insights from the social sciences. *Artificial Intelligence, 267*, 1–38. https://doi.org/10.1016/j.artint.2018.07.007

Mittelstadt, B. D., Russel, C., & Wachter, S. (2019). Explaining explanations in AI. *Proceedings of the conference on fairness, accountability, and transparency* (pp. 279–288). https://doi.org/10.1145/3287560.3287574

Molnar, C. et al. (2022). General pitfalls of model-agnostic interpretation methods for machine learning models. In A. Holzinger, R. Goebel, R. Fong, T. Moon, K. R. Müller & W. Samek (Eds.), *xxAI—Beyond explainable AI. xxAI 2020*. Lecture Notes in Computer Science(LNCS), vol. 13200. Springer. https://doi.org/10.1007/978-3-031-04083-2_4

Musiał, M. (2019). *Enchanting robots. Intimacy, magic, and technology*. Springer.

Nagel, S. K., & Reiner, P. B. (2013). Autonomy support to foster individuals' flourishing. *The American Journal of Bioethics, 6*, 36–37.

Newton, C. (2024). *Google's AI search setback*. https://www.platformer.news/google-ai-overviews-eat-rocks-glue-pizza/ (last accessed May 31st 2024).

Nunes, I., & Jannach, D. (2017). A systematic review and taxonomy of explanations in decision support and recommender systems. *User Modeling and User-Adapted Interaction, 27*, 393–444.

Nyholm, S. (2018). The ethics of crashes with self-driving cars: A roadmap. *Philosophy Compass*. https://compass.onlinelibrary.wiley.com/doi/full/10.1111/phc3.12507 (last accessed May 31st 2024)

Olshin, B. (2019). *Lost knowledge: The concept of vanished technologies and other human histories*. (Vol. 16). Brill.

Papagni, G., & Koeszegi, S. A. (2021). Pragmatic approach to the intentional stance semantic, empirical and ethical considerations for the design of artificial agents. *Minds & Machines, 31*, 505–534. https://doi.org/10.1007/s11023-021-09567-6

Pasquale, F. (2015). *The black box society: The secret algorithms that control money and information*. Harvard University Press

Pavlidis, G. (2024). Unlocking the black box: Analysing the EU artificial intelligence act's framework for explainability in AI. *Law, Innovation and Technology, 16*(1), 293–308. https://doi.org/10.1080/17579961.2024.2313795

Penu, O. K. A., Boateng, R., & Owusu, A. (2021). Towards explainable AI(xAI): Determining the factors for firms' Adoption and use of xAI in Sub-Saharan Africa (2021). *AMCIS 2021 TREOs. 35*. https://aisel.aisnet.org/treos_amcis2021/35

Ploug, T., & Holm, S. (2020). The four dimensions of contestable AI diagnostics—A patient-centric approach to explainable AI. *Artificial Intelligence in Medicine, 107*, 101901.

Price II, W. N. (2015). Black-Box medicine. *Harvard Journal of Law & Technology, 28*(2).

Ribeiro, M., Singh, S., & Guestrin, C. (2016). *Why should I trust you? Explaining the predictions of any classifier*. arXiv. https://arxiv.org/pdf/1602.04938.pdf

Robbins, S. A. (2019). Misdirected principle with a catch: Explicability for AI. *Minds & Machines, 29*, 495–514. https://doi.org/10.1007/s11023-019-09509-3

Rosenthal-von der Pütten, A., & Bock, N. (2023). Seriously, what did one robot say to the other? Being left out from communication by robots causes feelings of social exclusion. *Human-Machine Communication, 6*, 117–134. https://doi.org/10.30658/hmc.6.7

Rudin, C. (2019). Stop explaining black box machine learning models for high stakes decisions and use interpretable models instead. *Nature Machine Intelligence., 1*(5), 206–215.

Rueda, J., Rodríguez, J. D., Jounou, I. P., Hortal-Carmona, J., Ausín, T., & Rodríguez-Arias, D. (2022). 'Just' accuracy? Procedural fairness demands explainability in AI-based medical resource allocations. In *AI & Society, 39*(3), 1–12.

Sand, M., Durán, J. M., & Jongsma, K. R. (2021). Responsibility beyond design: Physicians' requirements for ethical medical AI. *Bioethics*. https://doi.org/10.1111/bioe.12887

Selbst, A., & Powles, J. (2017). Meaningful information and the right to explanation. *International Data Privacy Law, 7*(4), 233–242. https://doi.org/10.1093/idpl/ipx022

Smart, A., James, L., Hutchinson, B., Wu, S., & Vallor, S. (2020). Why reliabilism is not enough: Epistemic and moral justification in machine learning. In *Proceedings of the AAAI/ACM conference on AI, ethics, and society* (pp. 372–377). https://doi.org/ https://doi.org/10.1145/3375627.3375866

Smith, G. (2014). *Standard deviations: Flawed assumptions, tortured data, and other ways to lie With statistics*. Overlook Press.

Sun, Z., Shen, S., Cao, S., Liu, H., Li, C., Shen, Y., Gan, C., Gui, L. Y., Wang, Y. X., Keutzer, K., & Darrell, T. (2023). *Aligning large multimodal models with factually augmented rlhf*. arXiv preprint arXiv:2309.14525

Steinhaus, H. (1954). Length, shape and area. *Colloquium Mathematicum, 3*, 1–13. https://doi.org/10.4064/cm-3-1-1-13

Theunissen, M., & Browning, J. (2022). Putting explainable AI in context: Institutional explanations for medical AI. *Ethics and Information Technology, 24*, 23. https://doi.org/10.1007/s10676-022-09649-8

Toader, A. (2019, November 11). *Auditability of AI systems—Brake or acceleration to innovation?* https://doi.org/10.2139/ssrn.3526222

Tomsett, R., Braines, D., Harborne, D., Preece, A., & Chakraborty, S. (2018). *Interpretable to whom? A role-based model for analyzing interpretable machine learning systems*. arXiv preprint arXiv:1806.07552

Tjoa, E., Khok, H. J., Chouhan, T., & Cuntai, G. (2021). *Improving deep neural network classification confidence using heatmap-based eXplainable AI*. arXiv preprint arXiv:2201.00009

Tutt, A. (2016). An FDA for algorithms. *Administrative Law Review, 69*(1), 83–123.

Ursin, F., Timmermann, C., & Steger, F. (2022). Explicability of artificial intelligence in radiology: Is a fifth bioethical principle conceptually necessary? *Bioethics, 36*(2), 143–153. https://doi.org/10.1111/bioe.12918

Vredenburgh, K. (2022). "The right to explanation. *In the Journal of Political Philosophy, 30*(2), 209–229.

Wachter, S., Mittelstadt, B., & Floridi, L. (2017). Why a right to explanation of automated decision-making does not exist in the general data protection regulation. *International Data Privacy Law, 7*(2), 76–99. https://doi.org/10.1093/idpl/ipx005

Wachter, S., Mittelstadt, B., & Russel, C. (2018). Counterfactual explanations without opening the black box: Automated decisions and the GDPR. *Harvard Journal of Law & Technology, 31*(2), 842–861. https://doi.org/10.2139/ssrn.3063289

Wadden, J. J. (2022). Defining the undefinable: The black box problem in healthcare artificial intelligence. *Journal of Medical Ethics, 48*, 764–768.

Wang, H., et al. (2023). Scientific discovery in the age of artificial intelligence. *Nature, 620*, 47–60.

Watson, J. C. (2021). *Expertise: A philosophical introduction*. Bloomsbury.

Wood, N. G. (2024). Explainable AI in the military domain. *Ethics and Information Technology, 26*, 29. https://doi.org/10.1007/s10676-024-09762-w

Zednik, C. (2019). *Solving the black box problem: A normative framework for explainable artificial intelligence*. arXiv. https://arxiv.org/abs/1903.04361

Zerilli, J. (2022). Explaining machine learning decisions. *Philosophy of Science, 89*(1), 1–19. https://doi.org/10.1017/psa.2021.13

Zerilli, J., Knott, A., Maclaurin, J., & Gavaghan, C. (2019). Transparency in algorithmic and human decision-making: Is there a double standard? *Philosophy & Technology, 32*(4), 661–683. https://doi.org/10.1007/s13347-018-0330-6

Zhang, Y., & Chen, X. (2020). Explainable recommendation: A survey and new perspectives. *Foundations and Trends in Information Retrieval, 14*(1), 1–101.

INDEX

© The Author(s), under exclusive license to Springer Nature
Switzerland AG 2024
H. Kempt, *(Un)explainable Technology*,
https://doi.org/10.1007/978-3-031-68098-4